SolidWorks for the Sheet Metal Guy
Course 3: Unfolding

Joe Bucalo
Neil Bucalo

www.SheetMetalGuy.com

SolidWorks for the Sheet Metal Guy
Course 3: Unfolding
Published by
Sheet Metal Guy, LLC
P O Box 498283
Cincinnati, OH 45249
www.SheetMetalGuy.com

Every effort has been made to ensure that all information contained within this book is complete and accurate. However, Sheet Metal Guy assumes no responsibility for the use of said information, nor any infringement of the intellectual property rights of third parties which would result from such use.

Please visit our website at www.SheetMetalGuy.com.

Manufactured in the United States of America

All technical illustrations and CAD models in this book were produced using SolidWorks 2007.

Windows™ is a trademark of Microsoft Corporation.
SolidWorks® is a registered trademark of SolidWorks Corporation.

ISBN 978-0-9795666-3-9

About the Authors

Joe Bucalo is the founder and President of Applied Production, Inc. He has over 30 years experience in the sheet metal industry, working with and developing software products to change the manufacturing world. When CAD was just beginning on the PC, Joe played a major role in the development of ProFold, the first truly automatic 3D sheet metal unfolding program. When most people were using a wall chart of bend deductions, Joe was promoting the use of the K-factor for more accurate flat patterns.

Joe later paved the way for graphics based sheet metal CAM when he introduced ProFab, which allows the direct transfer of geometry from CAD to CAM. ProFab was the first CAM program in the sheet metal industry to include an automatic tool selection routine.

Joe continues to work with clients to solve their design and manufacturing problems. He has a thorough knowledge of the most popular CAD programs and understands the issues faced by sheet metal manufacturers.

Neil Bucalo has a diverse background, including mechanical design engineering, CAD/CAM support and training, engineering consulting, web development, and technical writing. Neil started his career in support and training of the CADKEY software at Computer Aided Technology, Inc. He then moved forward as a Certified SolidWorks Support Technician.

Upon moving to the Cincinnati area, Neil joined Applied Production, Inc., a SolidWorks Solution Partner, where he has provided customer support and written several user training documents. He also created and served as Editor of the CKD Magazine, dedicated to users of the CADKEY software.

Neil is a CAD expert, having many years of experience using numerous CAD systems, including AutoCAD, CADKEY/KeyCreator, Solid Edge, and of course SolidWorks.

For the record, Neil is Joe's nephew.

Tell Us What You Think!

As the reader of this book, you are our most important critic. We value your opinion and want to know how we are doing, good or bad. If you feel we missed something or could have done a better job, let us know. Also, if there are other areas of SolidWorks you feel need more explanation, tell us. We may be able to help.

You can email us at **books@SheetMetalGuy.com** to let us know what you did or didn't like about this book – as well as what we can do to make our books better.

When you write, please be sure to include the book's title as well as your name and contact information. We will carefully review your comments and share them with those whom helped make this book possible.

Table of Contents

Introduction

Unfolding your sheet metal parts in SolidWorks is really very simple. Once you have created a proper sheet metal part, you simply click on the Flatten icon and it unfolds.

The real trick is getting the correct size flat pattern. SolidWorks provides tables to specify the material gauge thicknesses and the bend parameters. This greatly helps to ensure that you are using the same bend parameters for all your parts as well as all of your people are doing it the same way.

While we cannot just give you the magic numbers to get the correct flat pattern, we are attempting to offer you the methods to establish the numbers that do work for you. Again, SolidWorks includes a number of great features to help you in this area, but you still need to do the calculations yourself and prove them out in the shop to ensure that your parts will be good.

Conventions Used in this Book

It is assumed that you have a working knowledge of SolidWorks and the menu structure. You may want to open SolidWorks and in the Help menu, go through the Online Tutorial. In the first few chapters, we show the full CommandManager to help you learn what icons to select. Later chapters show only the icon to be selected. Dialog boxes, toolbars, and icons are shown in the book. When several icons appear in a dialog box, the one which you should select is circled in the picture in the book. A circle will not appear on your SolidWorks screen.

Setting the Toolbars to Match the Book

The CommandManager is a context-sensitive toolbar that dynamically updates based on the toolbar you want to access. By default, it has toolbars embedded in it based on the document type.

When you click a button in the control area, the CommandManager updates to show that toolbar. For example, if you click **Sketch** in the control area, the "Sketch" toolbar appears in the CommandManager.

Use the CommandManager to access toolbar icons in a central location and to save space for the graphics area.

To access the CommandManager, first open a new document. To do this, click the **New** icon in the "Standard" toolbar, or pull down the "File" menu and pick **New**.

The **New SolidWorks Document** dialog box appears.

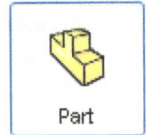

Click **Part** and then click **OK**. A new part window appears.

Pull down the "Tools" menu and pick **Customize**.

In the **Customize** dialog box on the **Toolbars** tab, make sure the **Enable CommandManager** check box is checked. If it is not, click the check box to check it. Then, select the **OK** button.

To make it easier for you to follow along and find the commands described in this book, you will want to make certain that the CommandManager is the same as displayed in the book.

To set up the CommandManager, move the cursor over the CommandManager and click the right mouse button. In the menu, make sure that **Show Description** is not checked. Then, pick **Customize Command Manager** as shown.

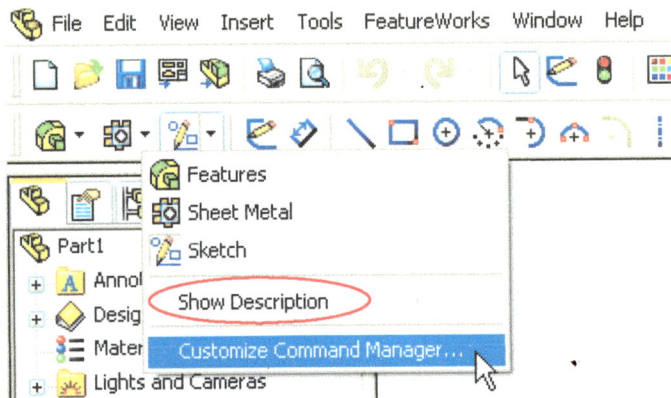

A long menu will appear, as shown to the right. Check **Features**, **Sheet Metal**, and **Sketch**. Make sure that all the others are unchecked. To accept your selections, simply click the left mouse button anywhere in the graphics area.

SolidWorks for the Sheet Metal Guy

Chapter 1

Material Types and Gauge Tables

The first step in creating a new part is to specify the known data. Material type and gauge are two easy and commonly known pieces of data for any sheet metal part. This ensures that the default values shown in the PropertyManager are the correct values for the part you are creating.

Once these are set, you can change them and any part details using the default values will update automatically.

This chapter shows you how to create a sheet metal gauge table and to specify the material type.

You can specify the sheet metal gauge table to be used from the PropertyManager while creating the base flange of your part. You then select one of the gauge values from the table to specify the material thickness for the part. You can create as many gauge tables as you would like. Make one for each material type you use which has its own gauge thickness, such as one for Aluminum and one for CRS. If you work with coated materials, you may want to make a separate table for this also. The sheet metal gauge tables store Gauge thickness, allowable bend radii, and the default k-factor.

If you change to a different table after your initial selection, or select a different gauge thickness, the default k-factor and bend radius are set to the default for the new selection. It is recommended that you edit key features to ensure that they have the correct inside radius and bend factor.

Making changes to the tables does not update your existing parts. For changes in the tables to take effect, you must open each part file, change to a different table and then change back to the edited table.

Define the Material Type

The first thing you want to specify when designing a new part is the type of material the part is to be made out of. SolidWorks allows you to create parts from a specific material by using the **Materials Editor PropertyManager**. The material and its properties propagate from SolidWorks to the part's mass properties, and material properties. The data is also used in the following third party applications: COSMOSWorks, COSMOSXpress, and PhotoWorks.

First, create a new **Part** document in SolidWorks by clicking the **New** icon in the "Standard" toolbar, or pull down the "File" menu and pick **New**.

By default, as you see in the FeatureManager design tree, the part material type is not specified.

In the FeatureManager design tree, right click on **Material**. Along with **Edit Material** the flyout menu also displays the ten most recently-used materials.

For now, pick **Edit Material** from the flyout menu. You may also pull down the "Edit" menu and pick **Appearance – Material**.

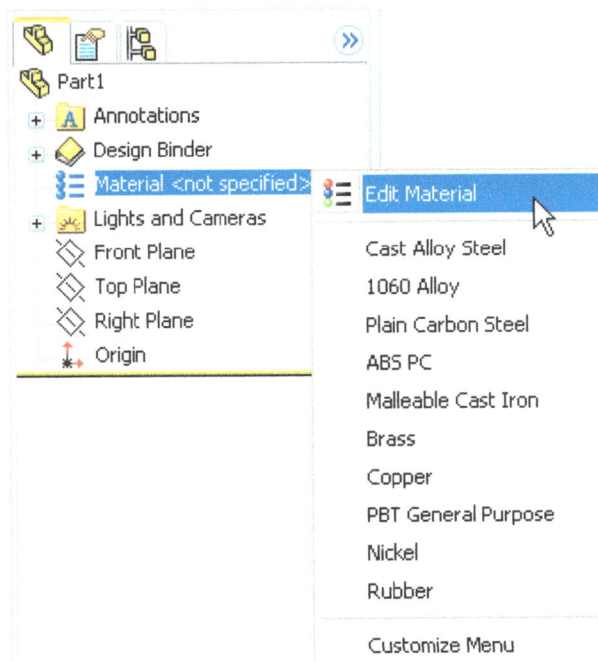

In the **Materials Editor** PropertyManager, SolidWorks provides an extensive list of common material types. Under **Materials**, click on the plus sign next to **Steel** to expand the list. Then, click on **Plain Carbon Steel**. You could have just as easily picked **Plain Carbon Steel** from the list.

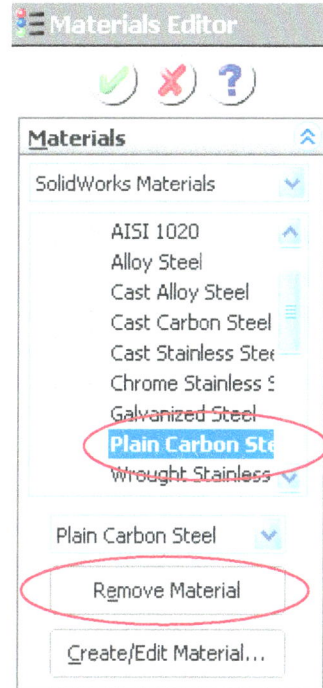

Click the green check mark button at the top of the **Materials Editor** PropertyManager.

In the FeatureManager design tree, the part material type is now **Plain Carbon Steel**.

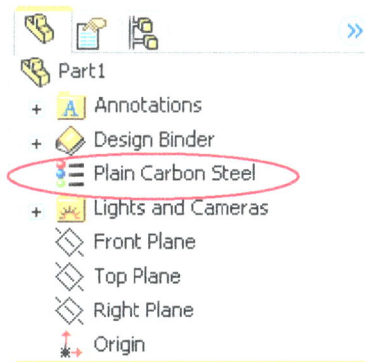

To remove the material from the part, right click on **Plain Carbon Steel** (the material) and **pick Edit Material**, or pull down the "Edit" menu and pick **Appearance – Material**.

In the PropertyManager, under **Materials**, click **Remove Material**,

Click the green check mark button at the top of the **Materials Editor** PropertyManager.

The material is removed from the part and the material name is removed from **Material** in the FeatureManager design tree.

Create a Custom Gauge Table

Sheet metal gauge tables store properties for selected materials. You can use the sheet metal gauge table to assign default values for the whole part. However, you don't want to create your own table from scratch. It is much easier to edit the sample table to ensure that everything will work properly. You will only need to set up your tables once. After they are set up correctly, you can forget about them.

To begin, you will copy a sample gauge table included in SolidWorks and modify the table to your own values.

To do so, open Microsoft Excel. Then, pull down the "File" menu and pick **Open**.

Browse to your default SolidWorks directory, also known as the <install_dir>. This is normally **C:\Program Files\SolidWorks**, specified when you installed your SolidWorks software. The sample tables are located in: **<install_dir>\lang\english\Sheet Metal Gauge Tables** folder. Open the folder named **lang**, then the folder named **english**, and finally the folder named **Sheet Metal Gauge Tables**.

Double click on the **sample table – steel – english units.xls** file.

Pull down the "File" menu and pick **Save As**.

In the **Save As** dialog box, change the file name to '**CRS Gauges.xls**' and press **Save**. It is a good idea to use the material types as file names, like CRS for Cold Rolled Steel. This will help you to easily find the gauge table that you need.

	A	B	C
1			
2	**Type:**	Steel Gauge Table	
3	**Process**	Steel Air Bending	
4	**K-Factor**	0.5	
5	**Unit:**	Inches	
6			
7	**Gauge No.**	**Gauge(Thickness)**	**Available Bend Radius**
8	3 Gauge	0.2391	.25; .50; .75
9	4 Gauge	0.2242	.25; .50; .75
10	5 Gauge	0.2092	.25; .50; .75
11	6 Gauge	0.1943	.20; .25; .50; .75
12	7 Gauge	0.1793	.20; .25; .50; .75
13	8 Gauge	0.1644	.20; .25; .50; .75
14	9 Gauge	0.1495	.15; .20; .25; .50
15	10 Gauge	0.1345	.15; .20; .25; .50
16	11 Gauge	0.1196	.125; .15; .20; .25; .50
17	12 Gauge	0.1046	.125; .15; .20; .25; .50
18	13 Gauge	0.0897	.10; .125; .15; .20; .25
19	14 Gauge	0.0747	.10; .125; .15; .20; .25
20	15 Gauge	0.0673	.075; .10; .125; .15; .20
21	16 Gauge	0.0598	.075; .10; .125; .15; .20
22	17 Gauge	0.0538	.075; .10; .125; .15; .20
23	18 Gauge	0.0478	.050; .075; .10; .125; .15

Leave the **Type** as **Steel Gauge Table** and the **Process** as **Steel Air Bending**. These are text values for labeling purposes. They are displayed but not really used.

Make sure that the **K-Factor** is set to '**0.3333**'. SolidWorks will use this value as the default K-factor used. If you leave K-factor blank, all the bend allowance values default to those last used.

Later, you will create a bend table to override this value.

Leave the **Units** set to **Inches**.

You will now edit the table to include 4 standard gauges; 10, 12, 16, and 20.

To do this, right click on row number **8** (3 Gauge) and pick **Delete**. Repeat the previous step to delete 4 Gauge to 9 Gauge.

Right click on row number **9** (11 Gauge) and pick **Delete**.

Right click on row number **10** (13 Gauge) and pick **Delete**. Repeat the previous step to delete 14 Gauge and 15 Gauge.

Right click on row number **11** (17 Gauge) and pick **Delete**.

	A	B	C
1			
2	**Type:**	Steel Gauge Table	
3	**Process**	Steel Air Bending	
4	**K-Factor**	0.3333	
5	**Unit:**	Inches	
6			
7	**Gauge No.**	**Gauge(Thickness)**	**Available Bend Radius**
8	10 Gauge	0.1345	.15; .20; .25; .50
9	12 Gauge	0.1046	.125; .15; .20; .25; .50
10	16 Gauge	0.0598	.075; .10; .125; .15; .20
11	18 Gauge	0.0478	.050; .075; .10; .125; .15

Right click on row **10** (16 Gauge) and pick **Cut**. Then, right click on row **8** (10 Gauge) and click **Insert Cut Cells**. The gauges may be listed in any order. The first gauge is the default used when you select the file as your gauge table. The gauge numbers do not have to be sequenced in order. The **Gauge No.** is just a text name. The gauge table values, columns B and C, are used by SolidWorks, but you can override these values.

Keep the existing values for the **Gauge(Thickness)** column. These are the actual thickness values which will be used.

The **Available Bend Radius** column lets you specify a list of the standard bend radii you utilize with each thickness. Click in cell **C8** and type '**.0625; .25; .50; .75**'. Make sure that you use a semi colon (;) as a delimiter between values. The numbers being used are make-believe numbers to show you an example. When you create your own table, enter the values you normally use in your manufacturing processes. In other words, if you only bend a 16th radius on 16 gauge material, then only show .0625 as the radius there. Don't put a radius there that you would never use.

Change the values to cell **C9**, **C10**, and **C11** as shown below. The number and value of **Available Bend Radius** can be unique for each Gauge No. These are the values that will be available in the **Base Flange** PropertyManager pull down list for the Bend Radius.

	A	B	C
1			
2	**Type:**	Steel Gauge Table	
3	**Process**	Steel Air Bending	
4	**K-Factor**	0.3333	
5	**Unit:**	Inches	
6			
7	**Gauge No.**	**Gauge(Thickness)**	**Available Bend Radius**
8	16 Gauge	0.0598	.0625; .25; .50; .75
9	10 Gauge	0.1345	.15; .25; .50; .75
10	12 Gauge	0.1046	.15; .25; .50
11	18 Gauge	0.0478	.050; .25; .50

In Excel, pull down the "File" menu and pick **Save**, or press **Ctrl+S**.

Create Another Custom Gauge Table

Pull down the "File" menu and pick **Save As**.

In the **Save As** dialog box, change the file name to '**Alum Gauges.xls**' and press **Save**. Make sure that the **Save in**: directory is still the **<install_dir>\lang\english\Sheet Metal Gauge Tables** folder.

Click in cell **B2** and type '**Aluminum**' to change the **Type**.

Click in cell **B3** and type '**Bending**' to change the **Process**.

Make sure that the **K-Factor** is set to '**0.33**'. Entering '1/3' will provide a more accurate number.

Leave the **Units** set to **Inches**.

Next, edit the table in column A to include the following **Gauge No.**: 1/4 in, 3/16 in, 10, 12, 14, and 16.

Then, edit the table in column B to include the following **Gauges**: 0.25, 0.1875, 0.1019, 0.0808, 0.0641, and 0.0508.

Finally, edit the table in column C to include the **Available Bend Radii** as shown below.

	A	B	C
1			
2	**Type:**	Aluminum	
3	**Process**	Bending	
4	**K-Factor**	0.33	
5	**Unit:**	Inches	
6			
7	**Gauge No.**	**Gauge(Thickness)**	**Available Bend Radius**
8	1/4 in	0.25	.125; .25
9	3/16 in	0.1875	.125; .25
10	10 Gauge	0.1019	.125; .25
11	12 Gauge	0.0808	.09; .125; .25
12	14 Gauge	0.0641	.09; .125; .25
13	16 Gauge	0.0508	.09; .125; .25

In Excel, pull down the "File" menu and pick **Save**, or press **Ctrl+S**.

Then, pull down the "File" menu and pick **Exit**.

Custom Gauge Table Locations

SolidWorks defaults to using the Sheet Metal Gauge Tables folder in the default installation directory. It is possible to add additional locations for your gauge tables. For example, if you are on a network and you use shared file locations, the same can be done for the gauge tables. This gets all of your users working from the same tables.

To do this, in SolidWorks, pull down the "Tools" menu and pick **Options**.

In the **System Options** dialog box, under the **System Options** tab, click on **File Locations**.

Pull down the **Show folder for**: menu and pick **Sheet Metal Gauge Table**.

The default location is shown. Click the **Add** button.

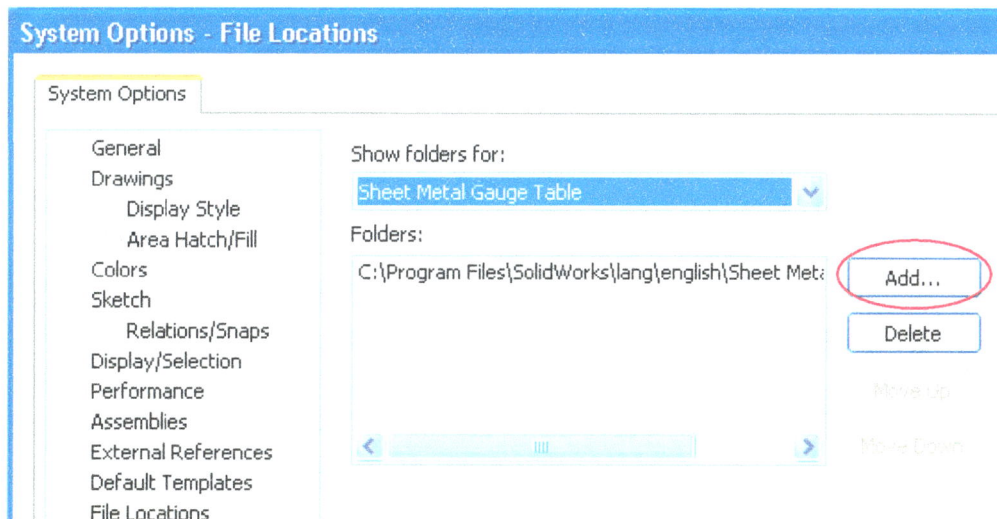

Select the appropriate network location in the **Browse For Folder** dialog box. You may want to create a new Sheet Metal Gauge Table folder on a network drive.

Once the desired location is highlighted, click **OK**.

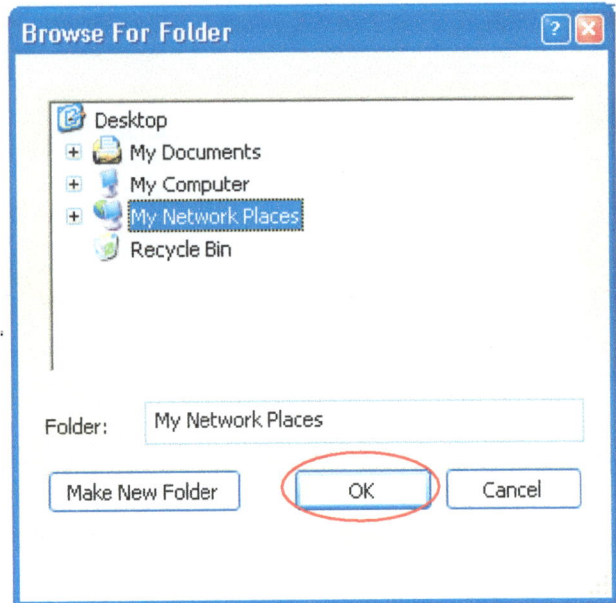

Create a Sheet Metal Part

Make sure that the part file is in inches.

To do this, pull down the "Tools" menu and pick **Options**.

On the **Document Properties** tab, select **Detailing**.

In the **Dimensioning standard** box, make sure that the standard is set to **ANSI**.

On the **Document Properties** tab, select **Units**.

In the **Unit system** box, select **IPS (inch, pound, second)**.

Click the **OK** button.

In the FeatureManager design tree, right click on **Material**, and pick **Plain Carbon Steel** from the menu.

Create a base flange by clicking the **Sheet Metal** icon in the control area of the CommandManager. Then, click the **Base-Flange/Tab** icon from the toolbar, or pull down the "Insert" menu and pick **Sheet Metal – Base Flange**.

Select the **Front** plane when prompted to select a plane on which to sketch the feature cross-section.

Create the sketch below using the **Line** icon in the CommandManager, or pull down the "Tools" menu and pick **Sketch Entities – Line**.

Press the **Escape** key.

Ctrl select the two vertical lines. In the **Properties** PropertyManager, under **Add Relations**, click the **Equal** button.

Add a '2' horizontal dimension to the bottom line and a '3' vertical dimension to the left vertical line using the **Smart Dimension** icon in the CommandManager, or pull down the "Tools" menu and pick **Dimensions – Smart**.

Exit the sketch by clicking the **Exit Sketch** icon in the CommandManager or in the upper right corner of the graphics area.

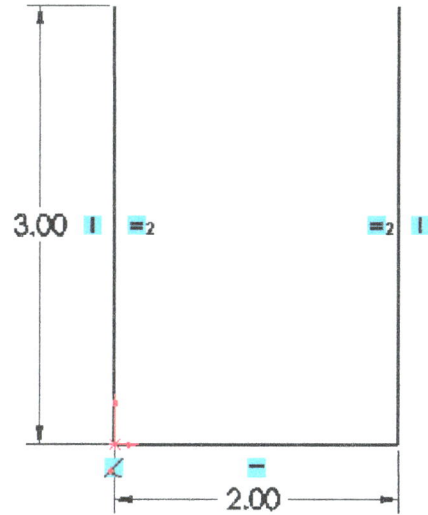

In the **Base Flange** PropertyManager, under **Direction 1**, set the **End Condition** to **Blind** and the **Depth** to '1'.

Under **Sheet Metal Gauges**, check the **Use gauge table** check box.

Pick **CRS GAUGES** from the **Select Table** pull down. Remember that the name of your file is the name of the gauge table shown in the menu. Make sure that you use a file name that you can easily identify the material type.

Under **Sheet Metal Parameters**, set the thickness to **16 Gauge**. Remember, the first gauge in your table is the default, or top, value used when you select the gauge table from the menu.

Check the **Reverse direction** check box, so the material is inside of the lines you drew.

Pull down the **Bend Radius** menu and pick **0.25in**. The **Bend Radius** menu uses the values that you specified for the gauge number in the gauge table created earlier.

Click the green check mark button at the top of the **Base Flange** PropertyManager to accept the settings and create the part.

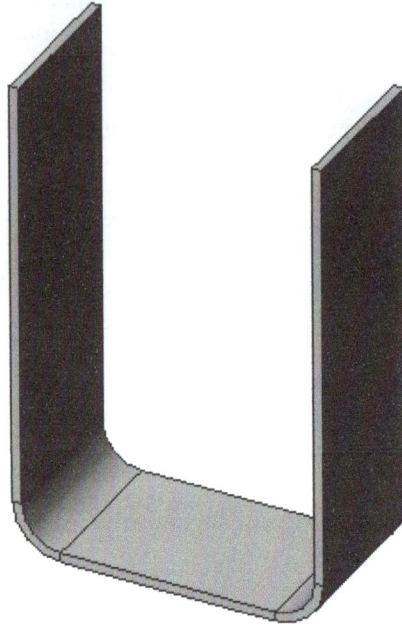

After creating the base flange, right click on **Sheet-Metal** in the FeatureManager design tree and pick **Edit Feature** to access the sheet metal gauge table and the **Sheet Metal Parameters**.

Click the green check mark button at the top of the **Sheet-Metal1** PropertyManager to return to the part.

Save the Part

Click the **Save** icon on the "Standard" toolbar, or pull down the "File" menu and pick **Save**.

The **Save As** dialog box appears. Make certain that you are in the desired folder. In the **File name** box, type '**U-Bracket**' and click **Save**.

Closing the File

Pull down the "File" menu and pick **Close**.

Click **No** when prompted to **Save changes to U-Bracket.SLDPRT?** to not save your changes.

Chapter 2

Bend Tables

The SolidWorks Bend tables make it easy for you to incorporate your bending information into your part designs. Whether it is a chart of Bend Allowance, Bend Deduction or a k-factor table, you can enter your data into an Excel spreadsheet.

A bend table differs from a gauge table in that a gauge table specifies the thicknesses and Radii available along with a single k-factor. A bend table provides specific bend data based on the material thickness, inside bend radius, and the angle of bend. When you incorporate a bend table, the bend table settings will override the K-factor defined in the gauge table.

There are some twists to this and it could require multiple Excel files to get the job done correctly. This chapter shows how to create a k-factor table, since it really is not documented by SolidWorks and can be confusing to figure out by yourself.

We have included samples of the Bend Allowance and Bend Deduction tables in the Appendix for your reference.

Create a Custom Bend Table

To begin, you will copy a sample bend table provided by SolidWorks and modify the table to be your own values. Again, you don't want to create your own table from scratch. You want to edit the sample table to ensure that everything will work properly.

An easy way to do this is to open Microsoft Excel. Then, in Excel, pull down the "File" menu and pick **Open**.

Browse to your default SolidWorks directory. The sample tables are located in: <install_dir>\lang\english\Sheetmetal Bend Tables\. Open the **lang** folder, then the **english** folder, and finally the **Sheetmetal Bend Tables** folder.

Double click on the **kfactor base bend table.xls** file.

Three different style tables are available. For this example, you will use a K-factor table. (See *Appendix A* for information about bend allowance and bend deduction tables.)

Pull down the "File" menu and pick **Save As**.

In the **Save As** dialog box, change the file name to '**SMG kfactor bend table.xls**' and click **Save**.

The **Type** is important because it tells SolidWorks what to do with the data in your chart. Make sure cell **B1** says '**K-Factor**', which ensures that the values entered in the table will be calculated as k-factor values.

Click in cell **B2** to highlight it. Then, type '**CRS**' and press **Enter**. The **Material** is a text value for labeling purposes. The values in the table are the important data used.

Enter the values in the Bend table, as shown below. The numbers used are make believe numbers to show you an example. You should build your table with the real values that you use in your manufacturing process. If you are unsure what numbers to use, see *Appendix B*.

When using the K-factor style of bend table (not the bend allowance or compensation tables), the values labeling the columns are ratios of radius divided by thickness and not values of thickness. Row 5 across the top of the table is these ratios. The table down the left side (**Column A**) is the different bend angles. Because the K-factor is unit-less, a ratio is used to keep the length unit-less as well. The values in the bend tables must be from lowest to highest (left to right). If an angle or ratio is used on a part which is not in the table, SolidWorks will interpolate values in between the values in the table to determine the k-factor, assuming that you have entered them in sequential order.

	A	B	C	D	E	F	G	H	I	J	K
1	Type:	K-Factor									
2	Material: CRS										
3											
4	Angle	Radius / Thickness									
5		0.25	0.50	0.75	1.00	1.25	1.50	2.00	2.50	3.00	5.00
6	15	0.50	0.50	0.50	0.50	0.50	0.50	0.50	0.50	0.50	0.50
7	30	0.45	0.48	0.50	0.50	0.50	0.50	0.50	0.50	0.50	0.50
8	45	0.40	0.43	0.45	0.46	0.50	0.50	0.50	0.50	0.50	0.50
9	60	0.33	0.38	0.40	0.42	0.46	0.50	0.50	0.50	0.50	0.50
10	75	0.28	0.33	0.35	0.38	0.42	0.46	0.50	0.50	0.50	0.50
11	90	0.25	0.28	0.31	0.33	0.38	0.42	0.50	0.50	0.50	0.50
12	120	0.22	0.24	0.28	0.30	0.35	0.40	0.46	0.47	0.48	0.48
13	150	0.19	0.20	0.24	0.28	0.33	0.38	0.42	0.44	0.45	0.48
14	180	0.15	0.17	0.20	0.25	0.30	0.35	0.38	0.40	0.42	0.45

Create similar charts for other materials that you use, like aluminum and brass. You can make as many of these tables as you like, one for each material type that you use, preferably the same ones as the material types. So, that everything relates back and forth.

In Excel, pull down the "File" menu and pick **Save**, or press **Ctrl+S**.

Then, pull down the "File" menu and pick **Exit**.

SolidWorks defaults to using the Sheet Metal Bend Tables folder in the default installation directory (it may even use the folder from a previous version, if you have previous versions installed). It is possible to add additional locations for your bend tables. For example, if you are on a network and you use shared file locations, the same can be done for the bend tables.

To do this, in SolidWorks, pull down the "Tools" menu and pick **Options**.

In the **System Options** dialog box, in the **System Options** tab, click on **File Locations**.

Pull down the "Show folder for:" menu and pick **Sheet Metal Bend Tables**.

The default location is shown. Click the **Add** button.

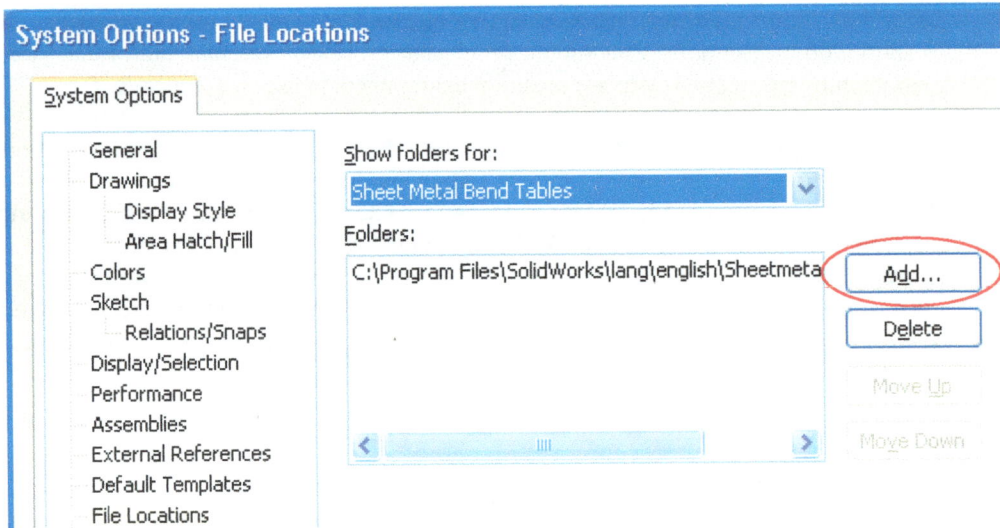

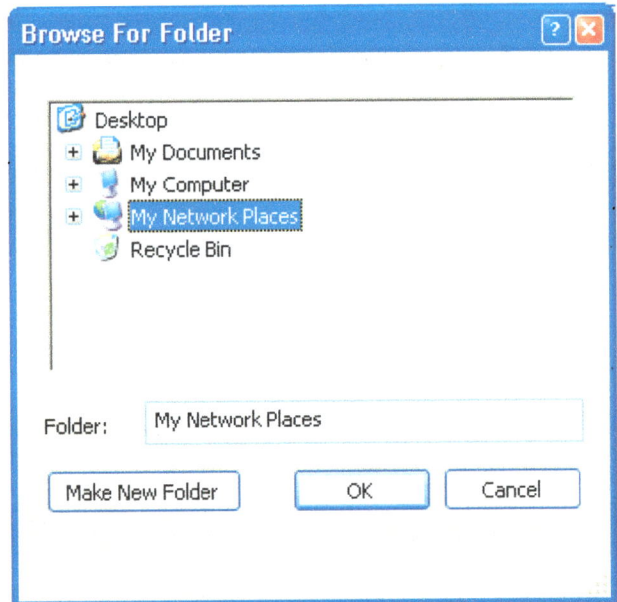

Select the appropriate network location in the **Browse For Folder** dialog box. You may want to create a new Sheet Metal Bend Table folder on a network drive.

Once the desired location is highlighted, click **OK**.

Create the Base Flange

You will now create a simple part to test the k-factor table and ensure that the values are being used correctly.

Begin a new **Part** document by clicking the **New** icon in the "Standard" toolbar, or pull down the "File" menu and pick **New**.

Create a base flange by clicking the **Sheet Metal** icon in the control area of the CommandManager. Then, click the **Base-Flange/Tab** icon from the toolbar, or pull down the "Insert" menu and pick **Sheet Metal – Base Flange**.

Select the **Front** plane when prompted to select a plane on which to sketch the feature cross-section.

Using the **Line** icon in the CommandManager, or pull down the "Tools" menu and pick **Sketch Entities – Line**, create a horizontal and vertical line starting at the origin as shown.

Then, add a '**2**' vertical dimension to the left vertical line and a '**3**' horizontal dimension to the bottom line using the **Smart Dimension** icon in the CommandManager, or pull down the "Tools" menu and pick **Dimensions – Smart**.

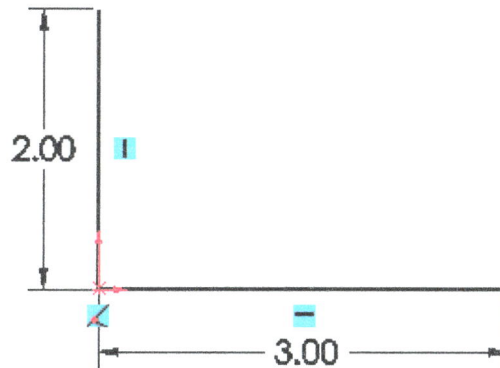

Exit the sketch by clicking the **Exit Sketch** icon in the CommandManager or in the upper right corner of the graphics area.

In the **Base Flange** PropertyManager, under **Direction 1**, set the **End Condition** to **Blind** and the **Depth** to '**1**'.

In the **Base Flange** PropertyManager, under **Sheet Metal Parameters**, set the **Thickness** to '**.125**'.

Make sure that the **Reverse direction** check box is checked.

Set the **Radius** to '**.0625**'. Take note that the ratio of the 0.125 thickness and the 0.0625 radius (0.125/0.0625) is 0.5.

Under **Bend Allowance**, pick **Bend Table** from the pull down menu. Then pick **SMG KFACTOR BASE BEND TABLE** from the **Bend Table** pull down menu.

Click the green check mark button at the top of the **Base Flange** PropertyManager to accept the settings and create the part.

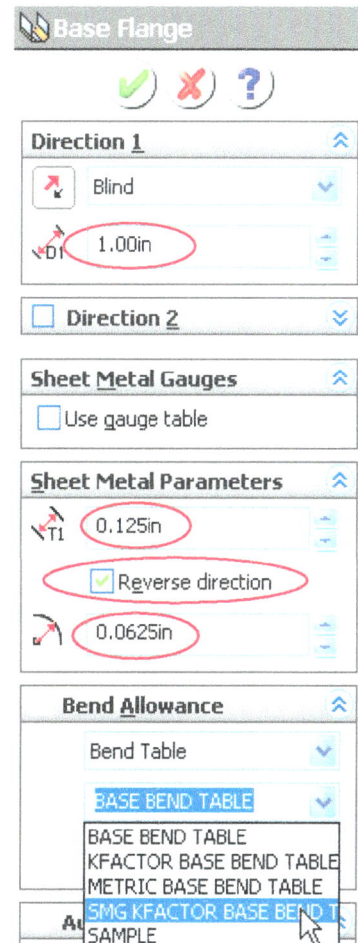

Verify the Bend Table Values

Click the **Flatten** icon from the toolbar. You may also right click on **Flat-Pattern1** in the FeatureManager design tree and select **Unsuppress**.

Press the **F** key on the keyboard to **Zoom to Fit**.

To verify the actual length of the flat part, you will measure the distance between the two end lines. To do this, pull down the "Tools" menu and pick **Measure**.

Click on the front edge of the part, as shown.

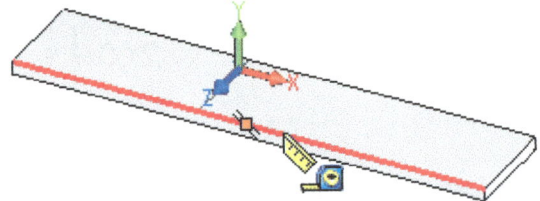

SolidWorks shows that the flat **Length** of the part is **4.78in**.

In the **Measure** dialog box, click on the **Units/Precision** icon.

The **Measure Units/Precision** dialog box appears.

Click on the **Use custom settings** radio button.

Then, change the **Decimal places** to '**4**'.

Click the **OK** button.

SolidWorks now shows that the **Length** is **4.7782in**.

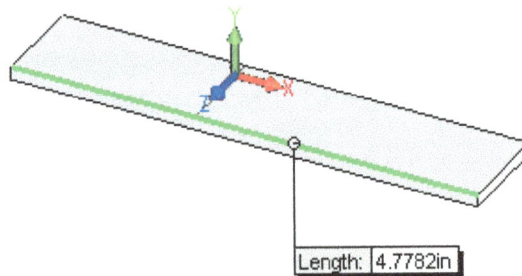

To verify that this is the correct size, add the original dimensions (2 + 3 = 5) and subtract it from the flat length measured above. This is the Outside Bend Compensation. So, 4.7782 − 5.00 = **-0.2218**.

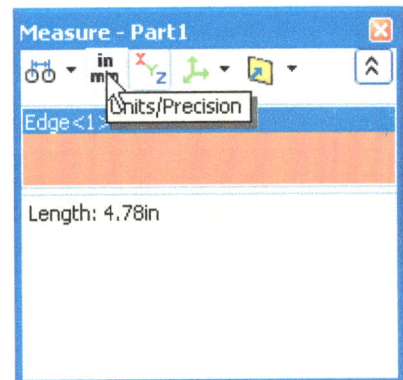

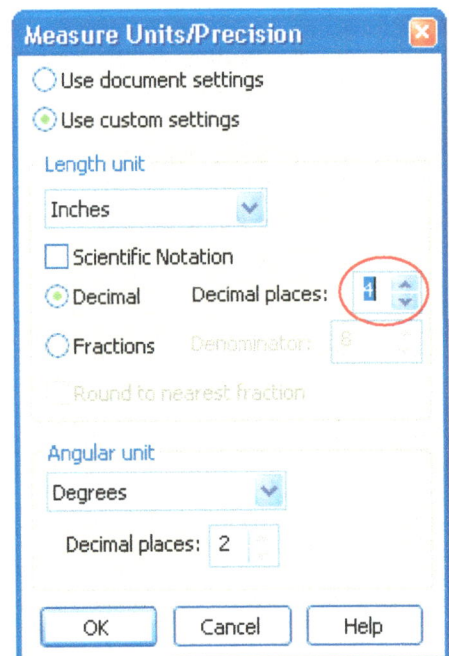

Verify with the Bend Calculator

Open your Internet Explorer and go to http://www.sheetmetalguy.com/

Click on the **Bend Calculator** page under **Valuable Tools**.

Click on the **Outside Comp** radio button and then enter '**-0.2218**' into the text field. Next, enter '**0.125**' for the **Material Thickness** and '**0.0625**' for the **Inside Radius**.

Click on the **Calculate** button.

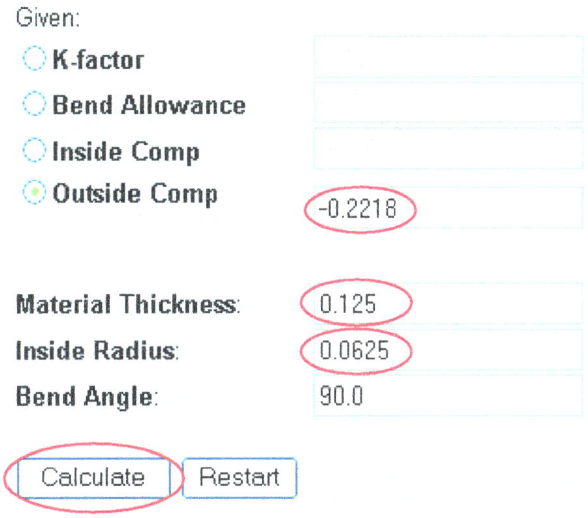

Given:

○ K-factor

○ Bend Allowance

○ Inside Comp

◉ Outside Comp   -0.2218

Material Thickness: 0.125

Inside Radius: 0.0625

Bend Angle: 90.0

[Calculate] [Restart]

The values are updated in the calculator. The **K-factor** value now displays **0.28**.

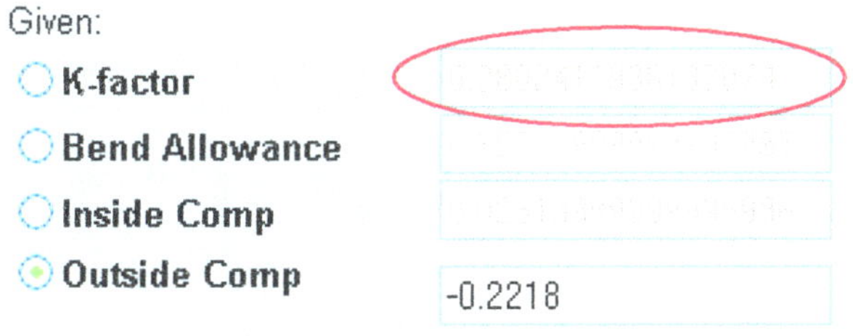

Given:

○ K-factor

○ Bend Allowance

○ Inside Comp

◉ Outside Comp   -0.2218

Close your Internet Explorer.

In SolidWorks, pull down the "Edit" menu and pick **Bend Table – Edit Table**.

The embedded bend table appears in your graphics area.

Calculate the ratio of Radius to Thickness by dividing the radius by the thickness (0.0625 / 0.125 = 0.5).

	A	B	C	D	E	F	G	H	I	J	K
1	Type:	K-Factor									
2	Material:	CRS									
3											
4	Angle					Radius / Thickness					
5		0.25	0.50	0.75	1.00	1.25	1.50	2.00	2.50	3.00	5.00
6	15	0.50	0.50	0.50	0.50	0.50	0.50	0.50	0.50	0.50	0.50
7	30	0.45	0.48	0.50	0.50	0.50	0.50	0.50	0.50	0.50	0.50
8	45	0.40	0.43	0.45	0.46	0.50	0.50	0.50	0.50	0.50	0.50
9	60	0.33	0.38	0.40	0.42	0.46	0.50	0.50	0.50	0.50	0.50
10	75	0.28	0.33	0.35	0.38	0.42	0.46	0.50	0.50	0.50	0.50
11	90	0.25	0.28	0.31	0.33	0.38	0.42	0.50	0.50	0.50	0.50
12	120	0.22	0.24	0.28	0.30	0.35	0.40	0.46	0.47	0.48	0.48
13	150	0.19	0.20	0.24	0.28	0.33	0.38	0.42	0.44	0.45	0.48
14	180	0.15	0.17	0.20	0.25	0.30	0.35	0.38	0.40	0.42	0.45

\Sheet1 / Sheet2 / Sheet3 /

Look at cell **C11**. With a **Radius / Thickness** ratio of **0.50** and an **Angle** of **90**, the K-factor in the table shows **0.28**. This is the same value which the **Bend Calculator** showed.

Click anywhere in the graphics area to close the bend table.

Save the Part

Click the **Save** icon in the "Standard" toolbar, or pick **Save** from the "File" pull down menu.

The **Save As** dialog box appears. In the **File name** box, type the name of the drawing number. For this chapter, use '**L-Bracket**' and click **Save**.

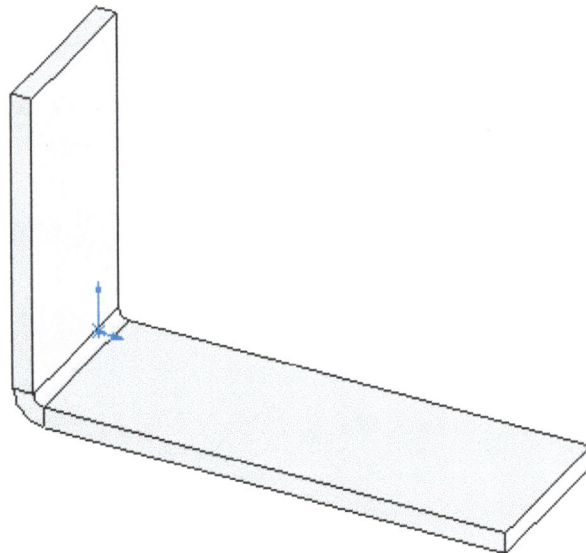

Chapter 3

Support Bracket

When constructing sheet metal parts with cutouts for pipes or tubes which are not perpendicular to the flanges, the original shape of the cutout is projected through the flange. SolidWorks can then create edge faces which are perpendicular to the flange or parallel to the direction of the cutout.

SolidWorks provides a check box for you to designate which method to use. When the Normal Cut check box is checked, the edges of the cutout are perpendicular to the face of the flange. This is the case in a punched or laser cut hole.

When working with thicker materials, it is desirable to taper the cutout so that the pipe or tube which passes through the cutout is a tighter fit. In this case, uncheck the box. This method would then require special machining to manufacture the cutout.

You will also see here, how to extend a cutout through multiple flanges, eliminating the need to place the same cutout on several flanges.

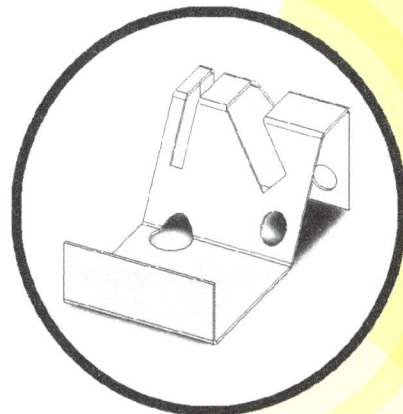

Create the Base Flange

Begin a new **Part** document by clicking the **New** icon in the "Standard" toolbar, or pull down the "File" menu and pick **New**.

In the FeatureManager design tree, right click on **Material**, and pick **Edit Material** from the menu. You may also pull down the "Edit" menu and pick **Appearance – Material**.

In the **Materials Editor** PropertyManager, under **Materials**, click on the plus sign next to **Aluminum Alloys** to expand the list. Then, click on **6061 Alloy**.

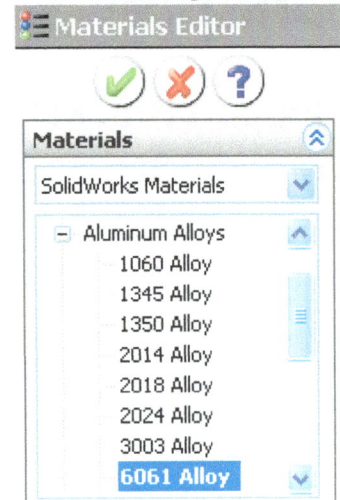

Click the green check mark button at the top of the **Materials Editor** PropertyManager.

In the FeatureManager design tree, the part material type is now **6061 Alloy**.

Create a base flange by clicking the **Sheet Metal** icon in the control area of the CommandManager. Then, click the **Base-Flange/Tab** icon from the toolbar, or pull down the "Insert" menu and pick **Sheet Metal – Base Flange**.

Select the **Right** plane when prompted to select a plane on which to sketch the feature cross-section.

Click the **Line** icon in the CommandManager, or pull down the "Tools" menu and pick **Sketch Entities – Line**.

Create a vertical line at the origin. Then, continue creating lines to create the sketch as shown below.

Click the **Smart Dimension** icon in the CommandManager, or pull down the "Tools" menu and pick **Dimensions – Smart**, and dimension the sketch as shown.

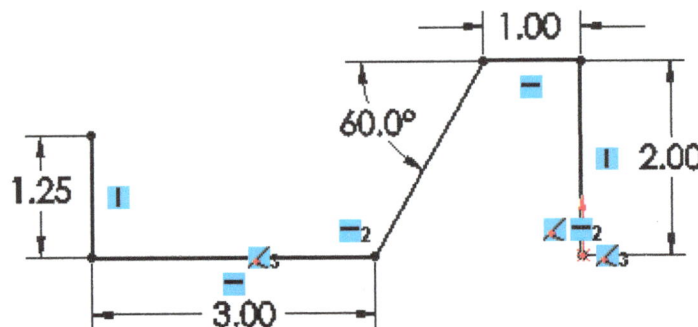

Exit the sketch by clicking the **Exit Sketch** icon in the CommandManager or in the upper right corner of the graphics area.

In the **Base Flange** PropertyManager, under **Direction 1**, set the **End Condition** to **Blind** and the **Depth** to '**3**'.

Under Sheet Metal Gauges, check the **Use gauge table** check box.

Pick **ALUM GAUGES** from the **Select Table** pull down.

Under **Sheet Metal Parameters**, set the thickness to **12 Gauge**.

Check the **Reverse direction** check box.

Pull down the **Bend Radius** menu and pick **0.13in**. Remember in Chapter 1, you entered this value as 0.125, but SolidWorks rounds it to 2 decimal places for display purposes only.

Pull down the **Bend Allowance Type** menu and pick **Bend table**.

Pull down the Bend Table menu and pick **SMG KFACTOR BEND TABLE**.

✓ Click the green check mark button at the top of the **Base Flange** PropertyManager to accept the settings and create the part.

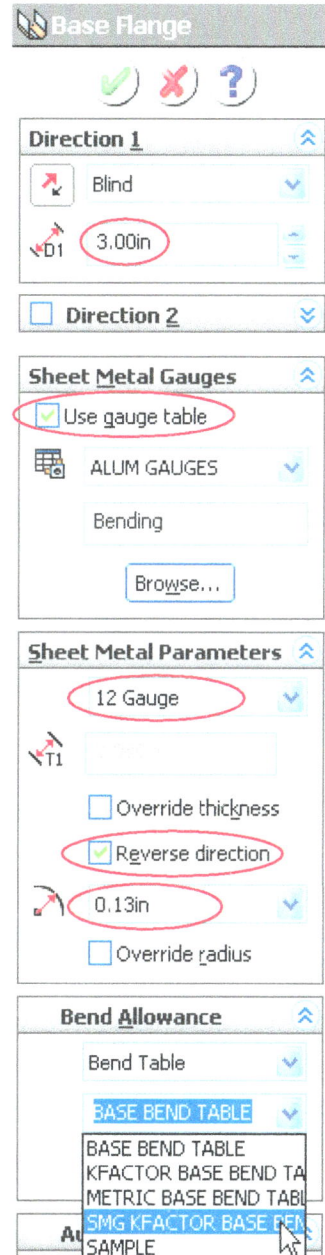

Create a Cut

In the bottom left corner of the graphics area, change the View orientation by clicking the pull down arrow and picking **Back**, or press **Ctrl+2**. | *Back | ▾ |

Click the **Extruded Cut** icon from the CommandManager or pull down the "Insert" menu and pick **Cut – Extrude**.

Select the back of the part as the plane onto which you will create the sketch.

⊕ Create a circle in the bottom left of the part using the **Circle** icon

in the CommandManager, or pull down the "Tools" menu and pick **Sketch Entities – Circle**.

Right click in the graphics area and pick **Smart Dimension** from the menu, and add the dimensions as shown.

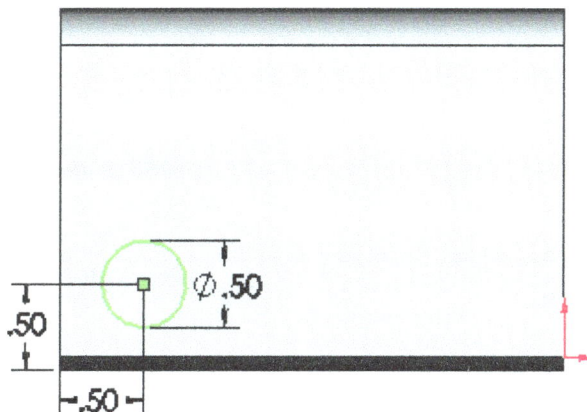

Click the **Line** icon in the CommandManager, or pull down the "Tools" menu and pick **Sketch Entities – Line**.

Create four perpendicular lines to create a rectangle at an angle as shown.

Right click in the graphics area and pick **Smart Dimension** from the menu, and add the dimensions as shown.

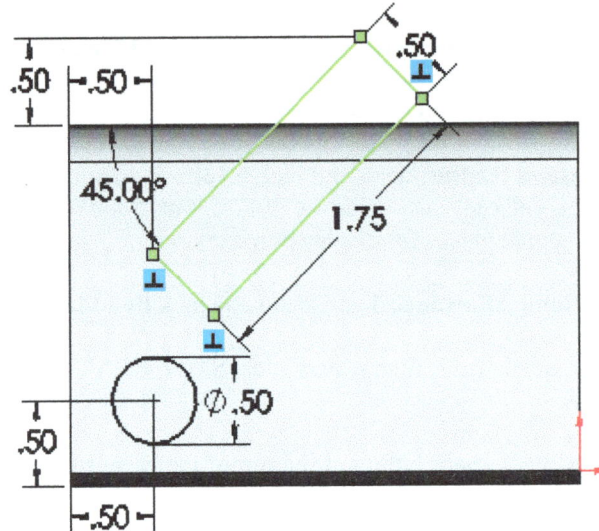

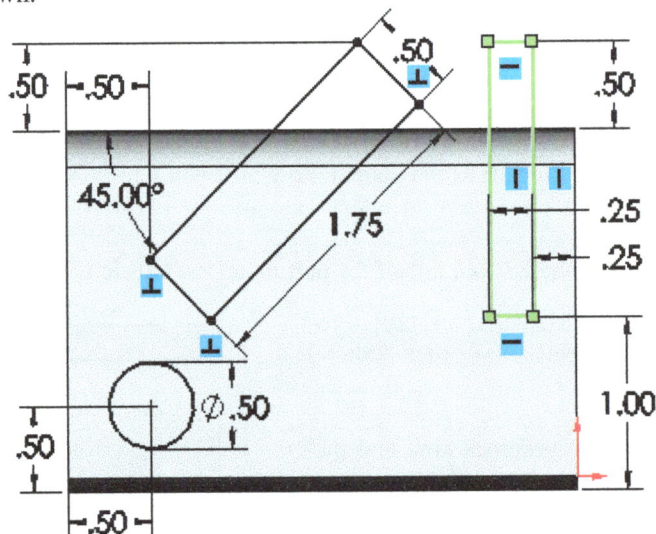

Create a rectangle on the right side of the sketch as shown using the **Rectangle** icon in the CommandManager, or pull down the "Tools" menu and pick **Sketch Entities – Rectangle**.

Right click in the graphics area and pick **Smart Dimension** from the menu, and add the dimensions as shown.

Exit the sketch by clicking the **Exit Sketch** icon in the CommandManager or in the upper right corner of the graphics area.

In the **Cut-Extrude** PropertyManager, under **Direction 1**, check **Link to thickness**, and make sure that **Normal cut** is checked.

Then, click the green check mark button at the top of the **Cut-Extrude** PropertyManager.

In the bottom left corner of the graphics area, change the View orientation by clicking the pull down arrow and picking **Trimetric**. `*Trimetric ▼`

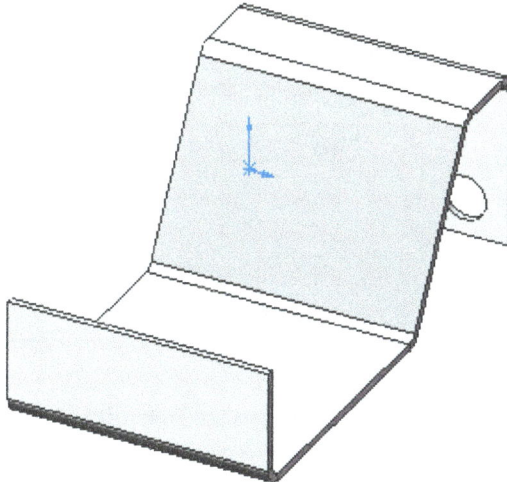

As you can see, the cut only goes through the back flange.

To make the cut go though the angled flange, in the FeatureManager design tree, right click on **Cut-Extrude1**, and pick **Edit Feature** from the menu.

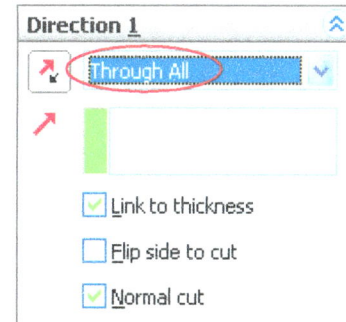

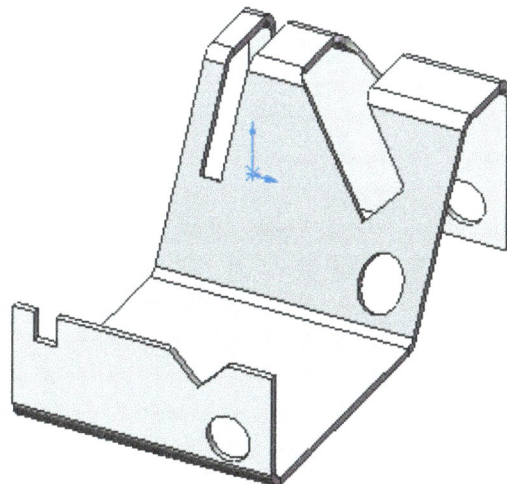

In the **Cut-Extrude1** PropertyManager, change the **End Condition** to **Through All**.

Click the green check mark button at the top of the **Cut-Extrude1** PropertyManager.

Now, the cut goes through the angled flange as intended, but it also comes through the front flange.

To fix this, in the FeatureManager design tree, right click on **Cut-Extrude1**, and pick **Edit Feature** from the menu.

In the **Cut-Extrude1** PropertyManager, change the **End Condition** to **Up to Surface**.

Then, click on the angled flange as shown.

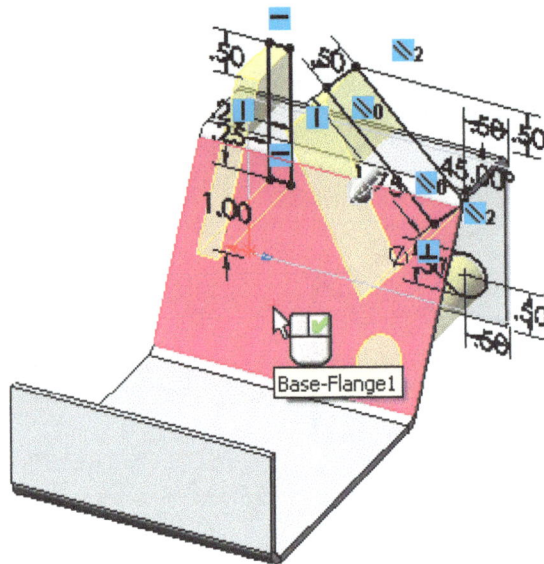

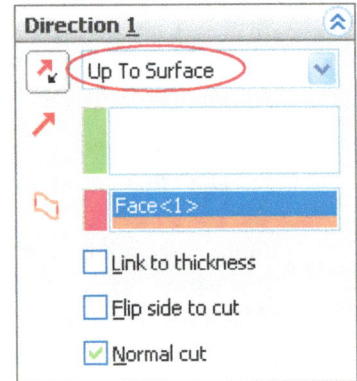

Direction 1

Up To Surface

Face<1>

☐ Link to thickness
☐ Flip side to cut
☑ Normal cut

✓ Finally, click the green check mark button at the top of the **Cut-Extrude1** PropertyManager.

Now, the cut goes through the back two flanges but not the front flange.

Create Another Cut

Click the **Extruded Cut** icon from the CommandManager or pull down the "Insert" menu and pick **Cut – Extrude**.

Select the top of the lower section of the part as the plane onto which you will create the sketch.

In the bottom left corner of the graphics area, change the View orientation by clicking the pull down arrow and picking **Top**, or press **Ctrl+5**.

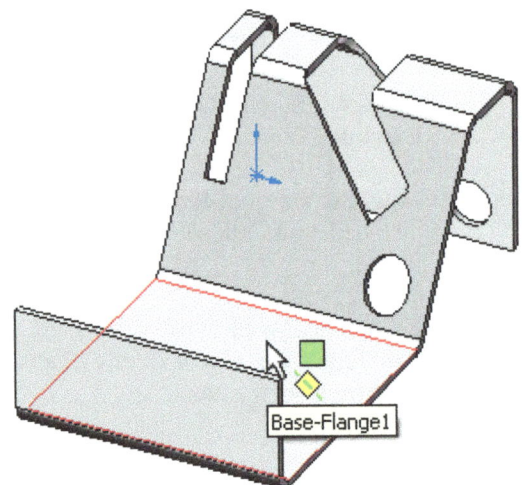

*Top ▾

SolidWorks for the Sheet Metal Guy

Create a circle in the bottom left of the part using the **Circle** icon in the CommandManager, or pull down the "Tools" menu and pick **Sketch Entities – Circle**.

Right click in the graphics area and pick **Smart Dimension** from the menu, and add the dimensions as shown.

Exit the sketch by clicking the **Exit Sketch** icon in the CommandManager or in the upper right corner of the graphics area.

In the **Cut-Extrude** PropertyManager, under **Direction 1**, check **Link to thickness**, and make sure that **Normal cut** is checked.

Then, click the green check mark button at the top of the **Cut-Extrude** PropertyManager.

In the bottom left corner of the graphics area, change the View orientation by clicking the pull down arrow and picking **Trimetric**.

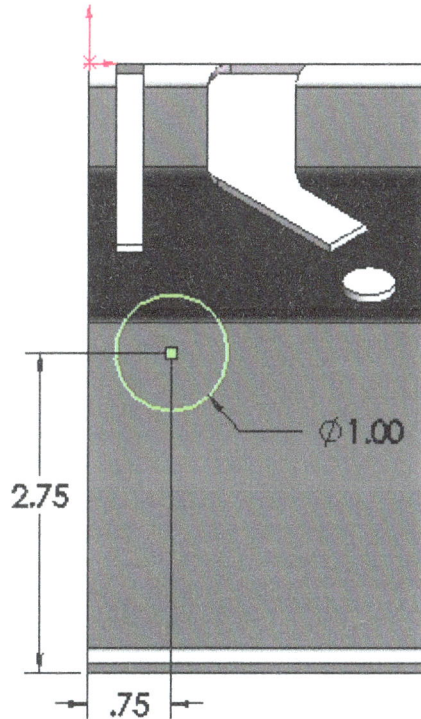

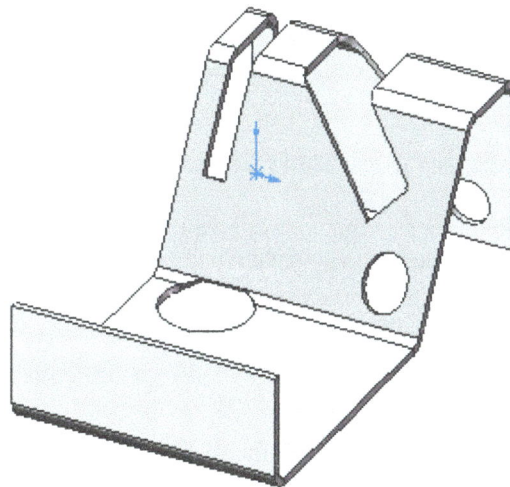

As you can see the cut does not come through the angled flange like it should.

To fix this, in the FeatureManager design tree, right click on **Cut-Extrude2**, and pick **Edit Feature** from the menu.

In the **Cut-Extrude2** PropertyManager, check **Direction 2**. Then, change the **End Condition** to **Up To Next**.

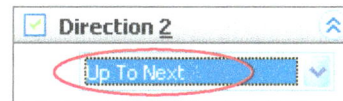

Click the green check mark button at the top of the **Cut-Extrude2** PropertyManager.

Now, the cut goes through the angled flange as intended.

The Flat Pattern

Click the **Sheet Metal** icon in the control area of the CommandManager. Then, click the **Flatten** icon from the toolbar. You may also right click on **Flat-Pattern1** and pick **Unsuppress**.

Click on the top of the flat pattern. Then, in the bottom left corner of the graphics area, click the pull down arrow and pick **Normal To**.

Notice the **1.00** diameter circle is a tear drop.

The bottom **0.50** circle is an oval.

The middle of the angled rectangle is wider than the ends.

Importantly, notice the extra notch in the corner.

This extra material is just a strange side effect of the normalization algorithm, which is not perfect. There are two ways you can think about an extrusion passing through the angled body and keeping uniform thickness. One way is to cut into the part as soon as the extrusion meets the inside face. But if you do that, the extrusion is not at the same location on the other side of the part. So, the other way is to cut into the part as soon as the extrusion meets the outside face. When **Normal to** is checked, SolidWorks does both. Normally, a notch is not seen. The notch that appears is caused by the angle of the sketch in combination of the angle of the flange.

This requires an extra step to get rid of this extra notch parametrically.

Click the **Sheet Metal** icon in the control area of the CommandManager. Then, click the **Flatten** icon from the toolbar. You may also right click on **Flat-Pattern1** and pick **Suppress**.

Click the **Extruded Cut** icon from the CommandManager or pull down the "Insert" menu and pick **Cut – Extrude**.

Select the front of the angled flange as the plane onto which you will create the sketch.

In the bottom left corner of the graphics area, change the View orientation by clicking the pull down arrow and picking **Normal To**.

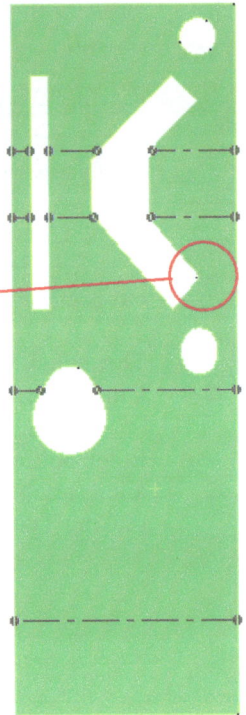

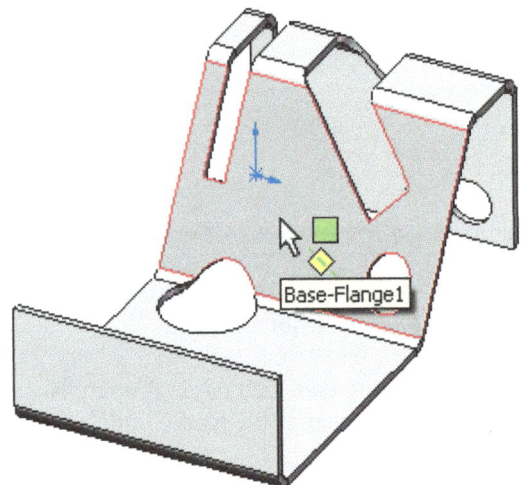

Ctrl select the three lines of the angled rectangle as shown.

Then, click the **Convert Entities** icon from the CommandManager, or pull down the "Tools" menu and pick **Sketch Tools – Convert Entities**. Relations are automatically created so that if the original shape changes size or location, this new sketch will automatically update to get rid of the notch.

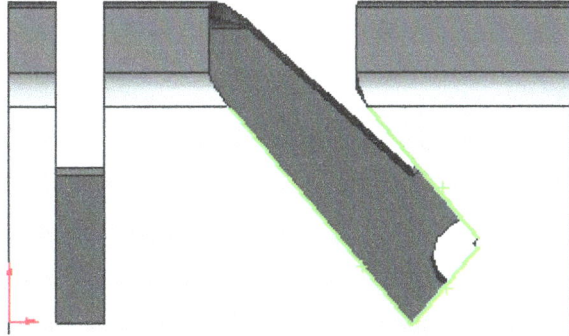

Create a line connecting the top endpoints of the shape using the **Line** icon in the CommandManager, or pull down the "Tools" menu and pick **Sketch Entities – Line**.

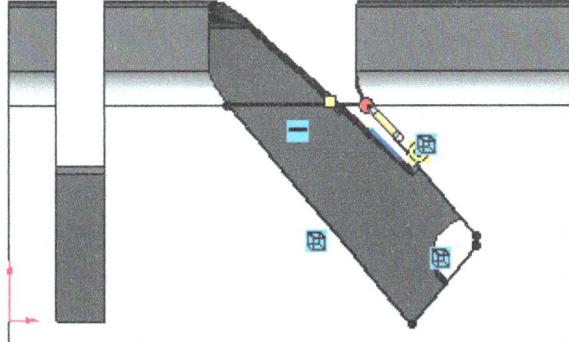

Press the **Escape** key to end the **Line** command.

In the lower right hand corner of the sketch is an open corner. Left click and drag the endpoints of the lines so that they intersect.

Then, click on the **Trim Entities** icon from the CommandManager, or pull down the "Tools" menu and pick **Sketch Tools – Trim**.

With **Power Trim** selected, drag the cursor over the two lines to trim them to a corner.

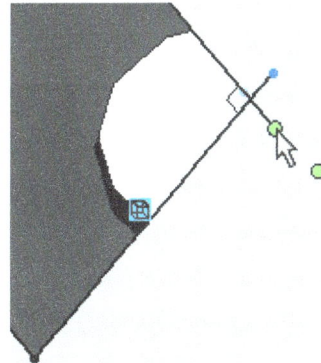

Exit the sketch by clicking the **Exit Sketch** icon in the CommandManager or in the upper right corner of the graphics area.

In the **Cut-Extrude** PropertyManager, under **Direction 1**, set the **End Condition** to **Blind**, check **Link to thickness**, make sure that **Normal cut** is checked, and make sure that **Direction 2** is unchecked.

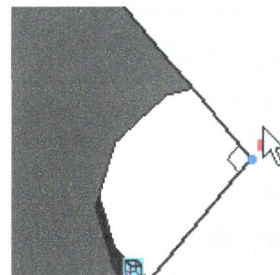

Then, click the green check mark button at the top of the **Cut-Extrude** PropertyManager.

In the bottom left corner of the graphics area, change the View orientation by clicking the pull down arrow and picking **Trimetric**.

`*Trimetric` ▾

Saving the Part

Click the **Save** icon in the "Standard" toolbar, or pick **Save** from the "File" pull down menu. In the **File name** box, type '**Support Bracket**' and select **Save**.

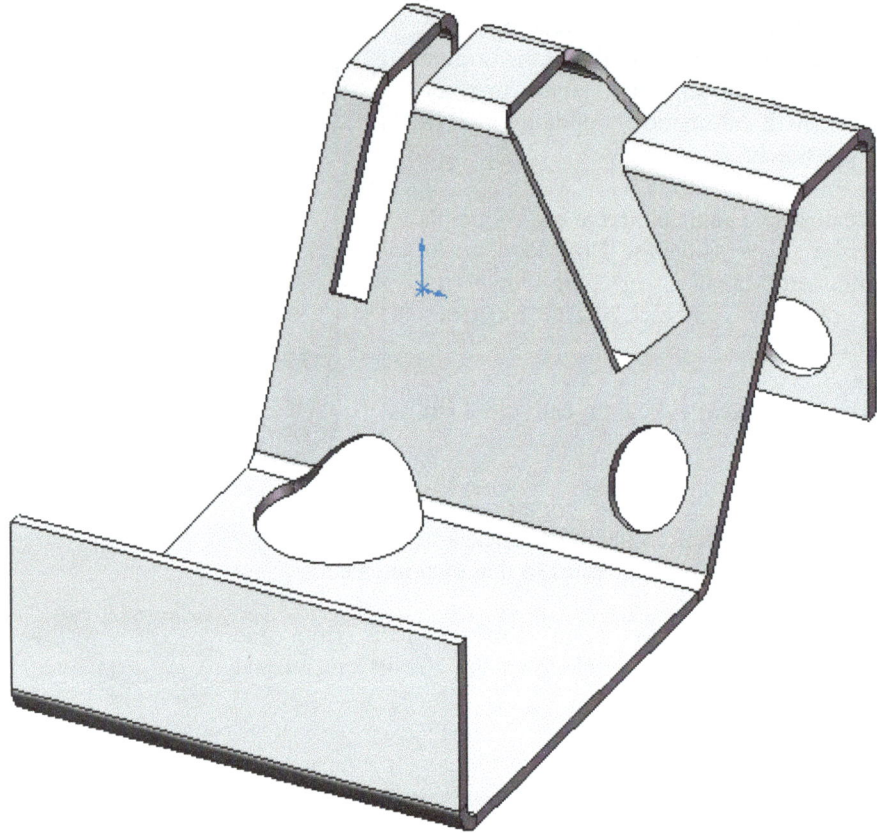

Chapter 4

Unfold and Fold

When a hole or feature is placed across the folded bend line, it causes the feature to be stretched. This most likely does not give you the result you wanted to achieve.

Even creating a single cut transposed onto two flanges can give the wrong result. Here, we look at using the Unfold and Fold features to add the hole in the flat and verify the results in the folded model.

The jog command was used here for convenience. The same problem exists with Edge flanges and Miter flanges. They are just more difficult to draw what you think you want in the folded model. So remember, always Unfold, add the feature, and Fold.

Create the Base Flange

Begin a new **Part** document by clicking the **New** icon in the "Standard" toolbar, or pull down the "File" menu and pick **New**.

In the FeatureManager design tree, right click on **Material**, and pick **Plain Carbon Steel** from the menu.

Create a base flange by clicking the **Sheet Metal** icon in the control area of the CommandManager. Then, click the **Base-Flange/Tab** icon from the toolbar, or pull down the "Insert" menu and pick **Sheet Metal – Base Flange**.

Select the **Top** plane when prompted to select a plane on which to sketch the feature cross-section.

Create a rectangle with the lower left corner at the origin using the **Rectangle** icon in the CommandManager, or pull down the "Tools" menu and pick **Sketch Entities – Rectangle**.

Click the **Smart Dimension** icon in the CommandManager, or pull down the "Tools" menu and pick **Dimensions – Smart**.

Add a '**17.75**' vertical dimension to the left vertical line and a '**7**' horizontal dimension to the bottom horizontal line.

Exit the sketch by clicking the **Exit Sketch** icon in the CommandManager or in the upper right corner of the graphics area.

In the **Base Flange** PropertyManager, under **Sheet Metal Gauges**, check the **Use gauge table** check box.

Pick **CRS GAUGES** from the **Select Table** pull down.

Use the default **16 Gauge Thickness**.

Make sure that the **Reverse direction** check box is checked.

Under **Bend Allowance**, pick **Bend Table** from the pull down menu. Then pick **SMG KFACTOR BASE BEND TABLE** from the **Bend Table** pull down menu.

Click the green check mark button at the top of the **Base Flange** PropertyManager to accept the settings and create the part.

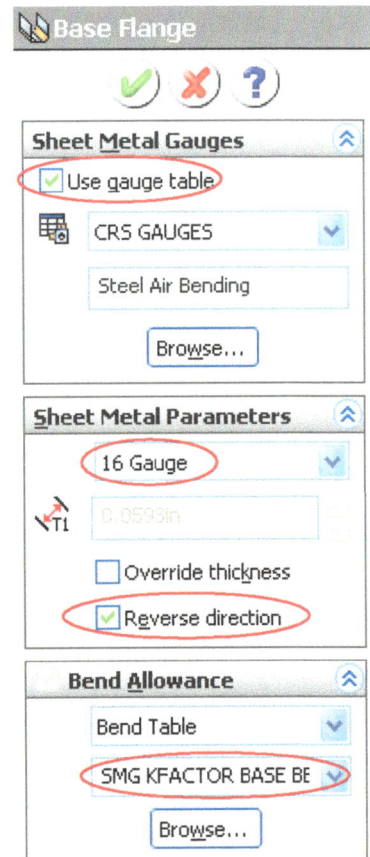

Base Flange

Sheet Metal Gauges
Use gauge table
CRS GAUGES
Steel Air Bending
Browse...

Sheet Metal Parameters
16 Gauge
0.0598in
Override thickness
Reverse direction

Bend Allowance
Bend Table
SMG KFACTOR BASE BE
Browse...

Create a Cut

Click the **Extruded Cut** icon from the CommandManager or pull down the "Insert" menu and pick **Cut – Extrude**.

Select the top of the part for the plane to sketch on.

In the bottom left corner of the graphics area, click the pull down arrow and pick **Normal To**.

Create a rectangle on the top of the part using the **Rectangle** icon in the CommandManager, or pull down the "Tools" menu and pick **Sketch Entities – Rectangle**.

Click the **Smart Dimension** icon in the CommandManager, or pull down the "Tools" menu and pick **Dimensions – Smart**, and add the two dimensions shown.

Pull down the "Tools" menu and pick **Sketch Tools – Rotate**.

Select the four lines that make up the rectangle.

Under **Center of rotation**, click in the **Base point** box, and select the bottom left corner of the rectangle.

Enter '**-40**' for the **Angle**.

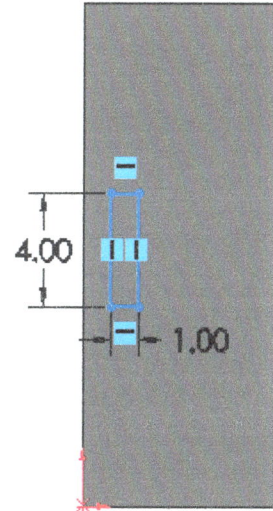

Click the green check mark button at the top of the **Rotate** PropertyManager.

Add two more dimensions to the left most corner of the sketch as shown using the **Smart Dimension** icon in the CommandManager, or pull down the "Tools" menu and pick **Dimensions – Smart**, and add the shown dimensions.

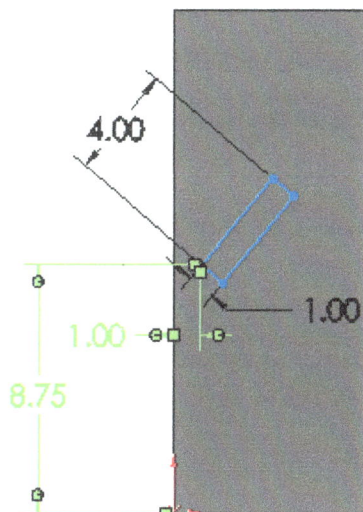

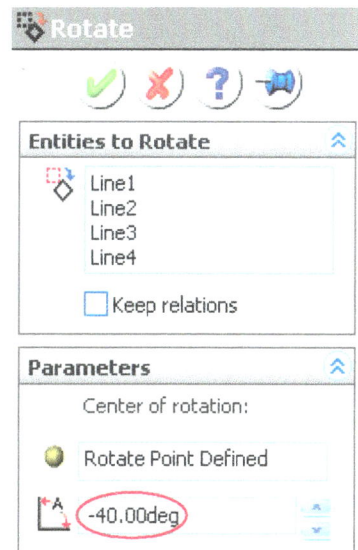

Exit the sketch by clicking the **Exit Sketch** icon in the CommandManager or in the upper right corner of the graphics area.

In the **Cut-Extrude** PropertyManager, check the **Link to thickness** check box and then click the green check mark button at the top of the **Cut-Extrude** PropertyManager.

Create a Jog

Click the **Jog** icon in the CommandManager, or pull down the "Insert" menu and pick **Sheet Metal – Jog**.

Select the top of the part.

Create a horizontal line across the cutout on the top of the part using the **Line** icon in the CommandManager, or pull down the "Tools" menu and pick **Sketch Entities – Line**.

Add an '**9.75**' dimension using the **Smart Dimension** icon in the CommandManager, or pull down the "Tools" menu and pick **Dimensions – Smart**.

Exit the sketch by clicking the **Exit Sketch** icon in the CommandManager or in the upper right corner of the graphics area.

In the bottom left corner of the graphics area, change the View orientation by clicking the pull down arrow and picking **Trimetric.** `*Trimetric  ▾`

In the graphics area, select the top face of the part below (left of) the horizontal line.

In the **Jog** PropertyManager, under **Jog Offset**, make sure that the **End Condition** is set to **Blind** and enter '**6**' for the **Offset Distance**.

Click the **Overall Dimension** button and the **Material Inside** button.

Make certain the **Jog Angle** is set to '**90**'.

Check the green check mark button at the top of the **Jog** PropertyManager to accept the settings and create the feature.

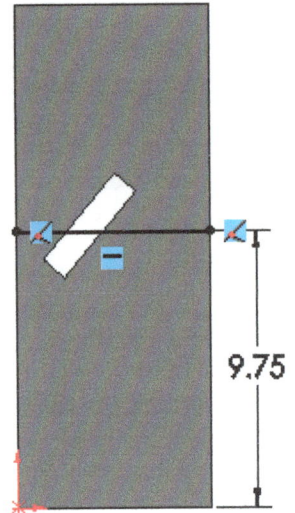

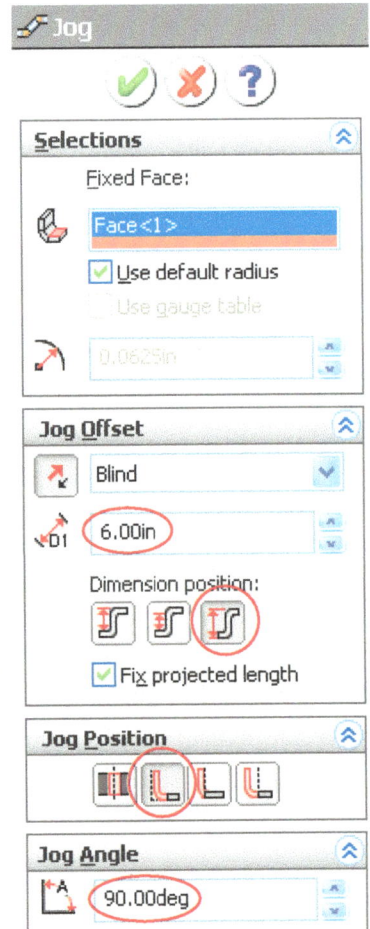

See how the cut is extended through the new jog flange.

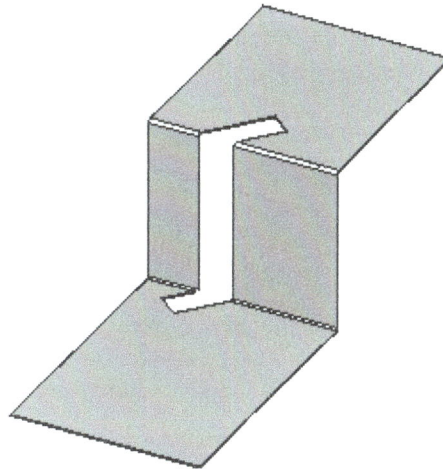

Click the **Flatten** icon in the CommandManager, or you may also right click on **Flat-Pattern1** in the FeatureManager design tree and pick **Unsuppress**.

In the bottom left corner of the graphics area, change the View orientation by clicking the pull down arrow and picking **Top**, or press **Ctrl+5**.

Notice the shape of the cutout.

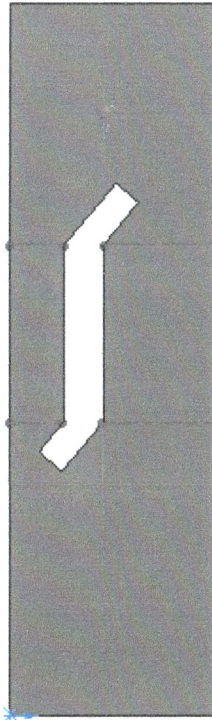

Click the **Flatten** icon in the CommandManager, or you may also right click on **Flat-Pattern1** in the FeatureManager design tree and pick **Suppress**.

Create the Correct Cut

The better way to do the cut is to do the cut in the flat pattern, like you would to punch it. Then, when the part is bent, the cut is displayed like it should be.

In the FeatureManager design tree, right click on **Cut-Extrude1** and pick **Suppress** from the menu.

In order to create the proper cut, you must first unbend the previous jog feature.

To do this, click the **Unfold** icon in the CommandManager, or pull down the "Insert" menu and pick **Sheet Metal – Unfold**.

For the **Fixed face**, select the bottom flat face of the part.

For the **Bends to unfold**, click the **Collect All Bends** button.

Click the green check mark button at the top of the **Unfold** PropertyManager and the bends will unfold.

Next, click the **Extruded Cut** icon from the CommandManager or pull down the "Insert" menu and pick **Cut – Extrude**.

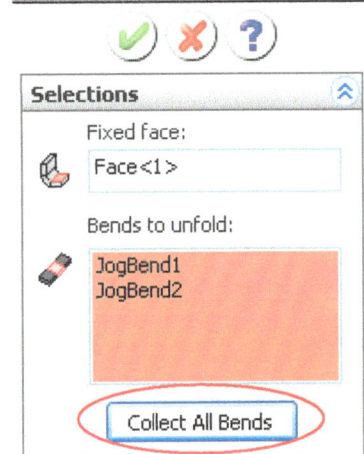

Select the top of the part for the plane to sketch on.

Create a rectangle on the top of the part using the **Rectangle** icon in the CommandManager, or pull down the "Tools" menu and pick **Sketch Entities – Rectangle**.

Click the **Smart Dimension** icon in the CommandManager, or pull down the "Tools" menu and pick **Dimensions – Smart**, and add the two dimensions shown.

Pull down the "Tools" menu and pick **Sketch Tools – Rotate**.

Select the four lines that make up the rectangle.

Under **Center of rotation**, click in the **Base point** box, and select the bottom left corner of the rectangle.

Enter '**-40**' for the **Angle**.

Click the green check mark button at the top of the **Rotate** PropertyManager.

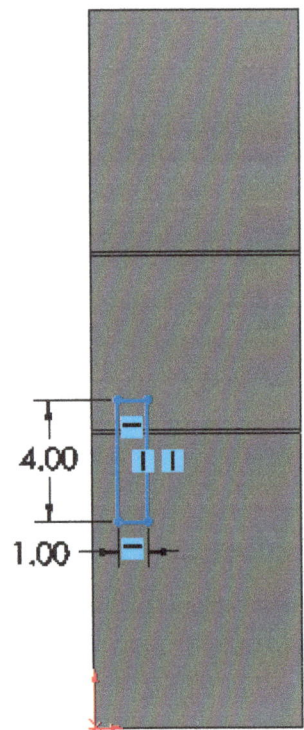

Add two more dimensions to the left most corner of the sketch as shown using the **Smart Dimension** icon in the CommandManager, or pull down the "Tools" menu and pick **Dimensions – Smart**.

Exit the sketch by clicking the **Exit Sketch** icon in the CommandManager or in the upper right corner of the graphics area.

In the **Cut-Extrude** PropertyManager, check the **Link to thickness** check box and then click the green check mark button at the top of the **Cut-Extrude** PropertyManager.

Change back to the **Trimetric** view by clicking the View orientation pull down arrow in the bottom left corner of your graphics area and picking **Trimetric**.

Click the **Fold** icon in the CommandManager, or pull down the "Insert" menu and pick **Sheet Metal – Fold**.

In the **Fold** PropertyManager, click on the **Collect All Bends** button to select all the appropriate bends in the part.

Click the green check mark button at the top of the **Fold** PropertyManager to fold the part back up.

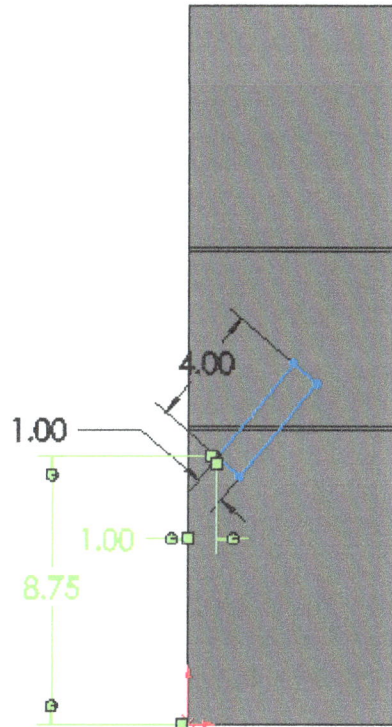

There is an extra sketch created when you fold up the part. In the graphics area, right click on the floating sketch and pick **Hide** from the menu.

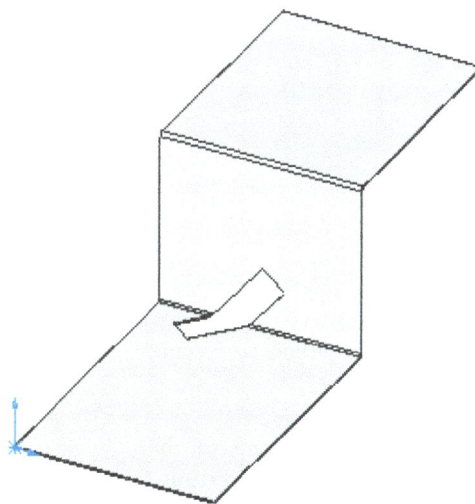

Click the **Flatten** icon in the CommandManager, or you may also right click on **Flat-Pattern1** in the FeatureManager design tree and pick **Unsuppress**.

In the bottom left corner of the graphics area, change the View orientation by clicking the pull down arrow and picking **Top**, or press **Ctrl+5**.

Notice the shape of the cutout. This is what we wanted.

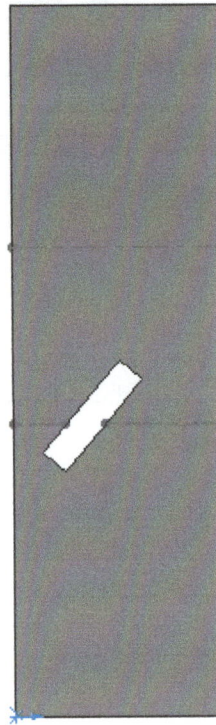

Click the **Flatten** icon in the CommandManager, or right click on **Flat-Pattern1** in the FeatureManager design tree and pick **Suppress**.

What About Holes?

In the FeatureManager design tree, right click on **Cut-Extrude2** and pick **Suppress** from the menu.

Right click on **Cut-Extrude1** and pick **Unsuppress** from the menu.

Then , right click on **Cut-Extrude1** again and pick **Edit Sketch** from the menu.

Create a circle using the **Circle** icon in the CommandManager, or pull down the "Tools" menu and pick **Sketch Entities – Circle**.

Add the dimensions as shown using the **Smart Dimension** icon in the CommandManager, or pull down the "Tools" menu and pick **Dimensions – Smart**.

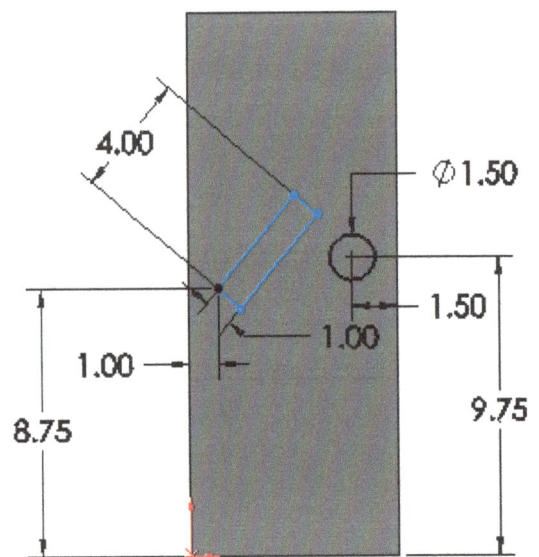

SolidWorks for the Sheet Metal Guy

Exit the sketch by clicking the **Exit Sketch** icon in the CommandManager or in the upper right corner of the graphics area.

If you get an error that starts as "Feature Jog1 failed to rebuild", then you need to edit the sketch related to the Jog feature and extend the bend line all the way across the part.

Click the **Flatten** icon in the CommandManager, or right click on **Flat-Pattern1** in the FeatureManager design tree and pick **Unsuppress**.

Notice the shape of the cutouts.

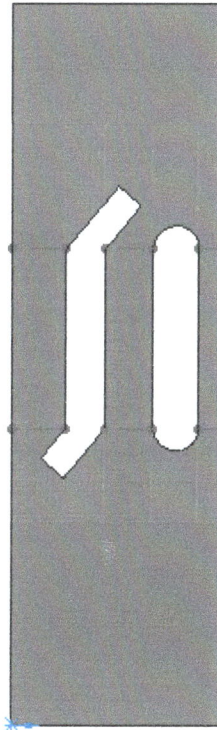

Click the **Flatten** icon in the CommandManager, or right click on **Flat-Pattern1** in the FeatureManager design tree and pick **Suppress**.

Click the View orientation pull down arrow in the bottom left corner of your graphics area and pick **Trimetric**.

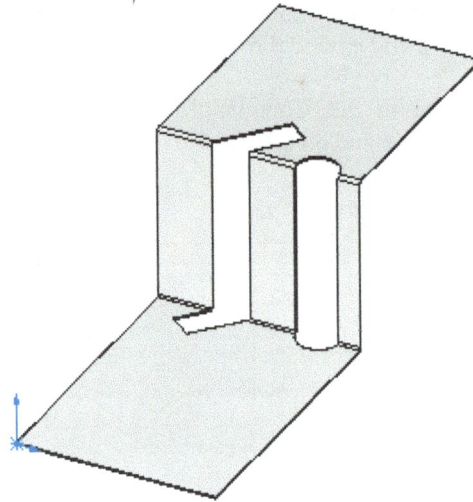

In the FeatureManager design tree, right click on **Cut-Extrude1** and pick **Suppress** from the menu.

Right click on **Cut-Extrude2** and pick **Unsuppress** from the menu.

Then, right click on **Cut-Extrude2** again and pick **Edit Sketch** from the menu.

In the bottom left corner of the graphics area, change the View orientation by clicking the pull down arrow and picking **Top**, or press **Ctrl+5**.

Create a circle using the **Circle** icon in the CommandManager, or pull down the "Tools" menu and pick **Sketch Entities – Circle**.

Add the dimensions as shown using the **Smart Dimension** icon in the CommandManager, or pull down the "Tools" menu and pick **Dimensions – Smart**.

Exit the sketch by clicking the **Exit Sketch** icon in the CommandManager or in the upper right corner of the graphics area.

Click the **Flatten** icon in the CommandManager, or right click on **Flat-Pattern1** in the FeatureManager design tree and pick **Unsuppress**.

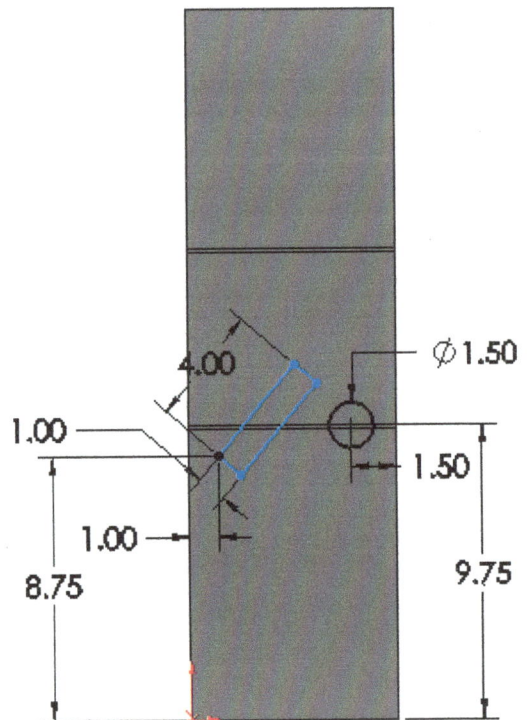

Notice the shape of the cutouts.

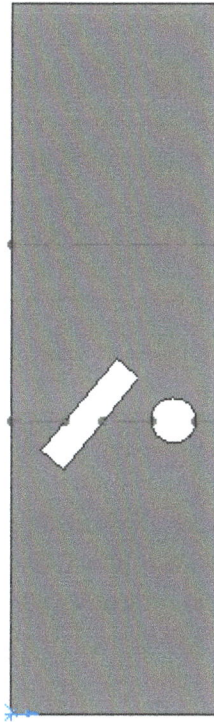

Click the **Flatten** icon in the CommandManager, or right click on **Flat-Pattern1** in the CommandManager and pick **Suppress**.

Click the View orientation pull down arrow in the bottom left corner of your graphics area and pick **Trimetric**.

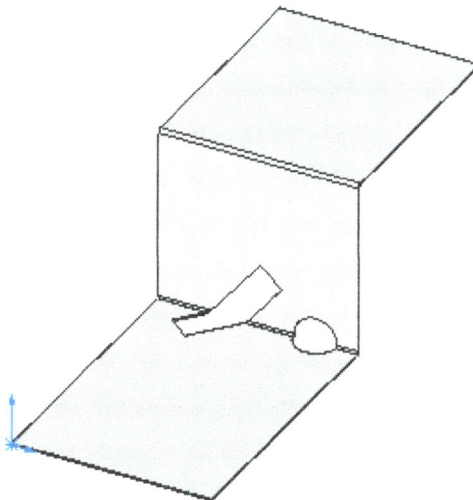

Save the Part

Click the **Save** icon in the "Standard" toolbar, or pick **Save** from the "File" pull down menu.

The **Save As** dialog box appears.

In the **File name** box, type the name of the drawing number. For this chapter, use '**Z-Bracket-4**' and select **Save**.

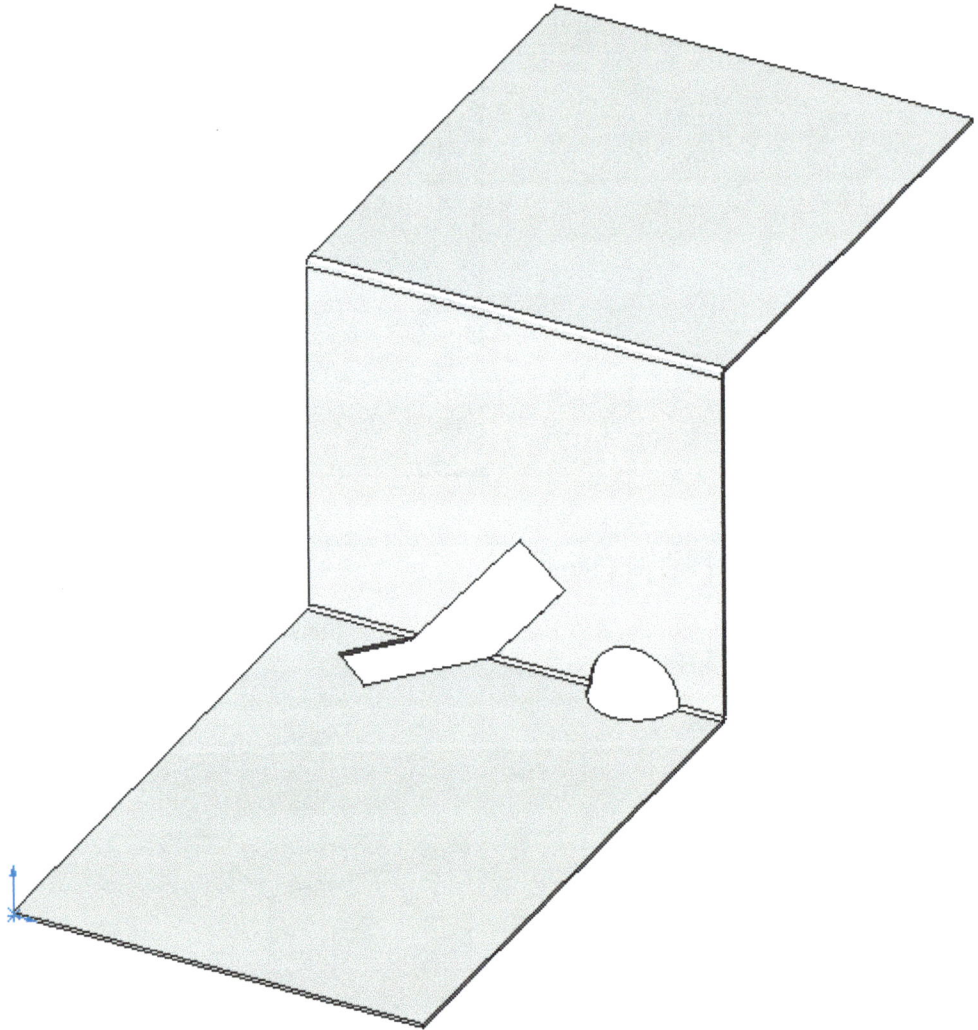

Chapter 5

Designing in the Flat

Here is a case where drawing in 3D just doesn't get it done. You know that the part will be made from a piece of bar stock (a rectangular blank). But the bend lines are not perpendicular to the edges of the stock.

When you draw what you think it should be in 3D, the flat is all wrong.

To show this, we will make the part in 3D and then as a flat and folded up. The base flange is the same in both parts, but the side flanges are very different.

In this chapter you will create the part both ways and compare the results.

Create a Part using the Edge Flange

☐ Begin a new **Part** document by clicking the **New** icon in the "Standard" toolbar, or pull down the "File" menu and pick **New**.

In the FeatureManager design tree, right click on **Material**, and pick **Plain Carbon Steel** from the menu.

🖐 Create a base flange by clicking the **Sheet Metal** icon in the control area of the CommandManager. Then, click the **Base-Flange/Tab** icon from the toolbar, or pull down the "Insert" menu and pick **Sheet Metal – Base Flange**.

Select the **Top** plane when prompted to select a plane.

✏ Starting at the origin, create the shape as shown below using the **Line** icon in the CommandManager, or pull down the "Tools" menu and pick **Sketch Entities – Line**.

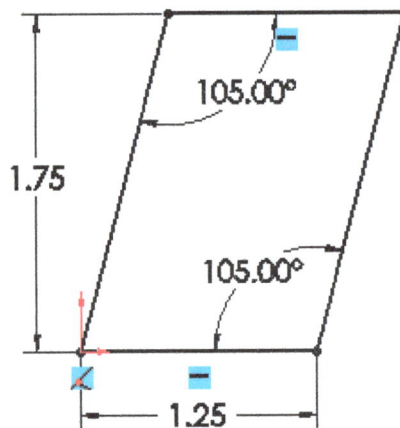

✐ Add the above dimensions using the **Smart Dimension** icon in the CommandManager, or pull down the "Tools" menu and pick **Dimensions – Smart**.

✍ Exit the sketch by clicking the **Exit Sketch** icon in the CommandManager or in the upper right corner of the graphics area.

In the **Base Flange** PropertyManager, under **Sheet Metal Gauges**, check the **Use gauge table** check box.

Pick **CRS GAUGES** from the **Select Table** pull down.

Use the default **16 Gauge Thickness**, and make sure that the **Reverse direction** check box is checked.

Under **Bend Allowance**, pick **Bend Table** from the pull down menu. Then pick **SMG KFACTOR BASE BEND TABLE** from the **Bend Table** pull down menu.

✅ Click the green check mark button at the top of the **Base Flange** PropertyManager to accept the settings and create the part.

Click the **Edge Flange** icon in the CommandManager, or pull down the "Insert" menu and pick **Sheet Metal – Edge Flange**.

Select the right edge of the base flange and move to the right and up and click to set the direction of the flange.

In the **Edge-Flange** PropertyManager, enter '**45**' for the **Angle**.

Set the **Flange Length** to **Blind** and the **Length** to '**2**'.

Click the **Outer Virtual Sharp** button and the **Bend Outside** button.

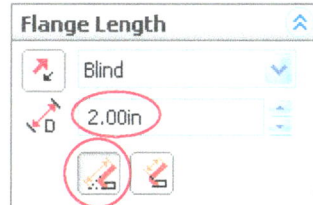

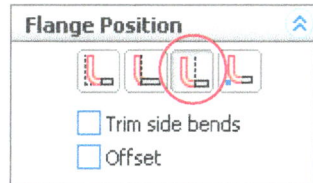

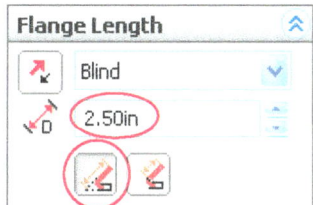

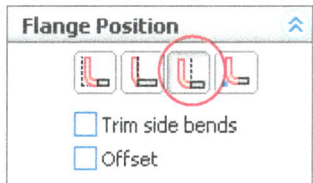

Click the green check mark button at the top of the **Edge-Flange** PropertyManager to accept the settings and create the flange.

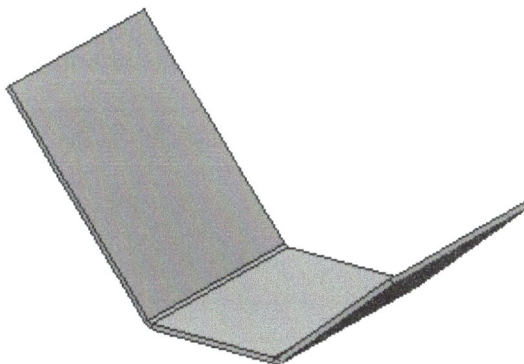

Click the **Edge Flange** icon in the CommandManager, or pull down the "Insert" menu and pick **Sheet Metal – Edge Flange**.

Select the left edge of the base flange and move to the left and up and click to set the direction of the flange.

In the **Edge-Flange** PropertyManager, enter '**45**' for the **Angle**.

Set the **Flange Length** to **Blind** and the **Length** to '**2.5**'.

Click the **Outer Virtual Sharp** button and the **Bend Outside** button.

Click the green check mark button at the top of the **Edge-Flange** PropertyManager to accept the settings and create the flange.

Angle

45.00deg

Flange Length

Blind

2.00in

Flange Position

Trim side bends

Offset

Angle

45.00deg

Flange Length

Blind

2.50in

Flange Position

Trim side bends

Offset

Click the **Flatten** icon in the CommandManager, or right click on **Flat-Pattern1** in the FeatureManager design tree and pick **Unsuppress**.

In the bottom left corner of the graphics area, change the View orientation by clicking the pull down arrow and picking **Top**, or press **Ctrl+5**.

The result here is a staggered edge part. Most likely this is not what you wanted to start with for the blank of the part. So, save this part and try it again starting with the flat pattern.

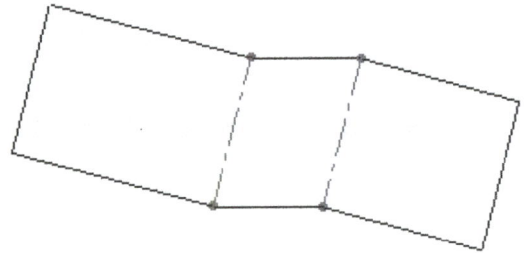

Save the Part

Click the **Save** icon in the "Standard" toolbar, or pick **Save** from the "File" pull down menu.

The **Save As** dialog box appears.

In the **File name** box, type the name of the drawing number. For this chapter, use '**Chapter 5 – Edge Flange**' and select **Save**.

Designing the Same Part in the Flat

Begin a new **Part** document by clicking the **New** icon in the "Standard" toolbar, or pull down the "File" menu and pick **New**.

In the FeatureManager design tree, right click on **Material**, and pick **Plain Carbon Steel** from the menu.

Create a base flange by clicking the **Sheet Metal** icon in the control area of the CommandManager. Then, click the **Base-Flange/Tab** icon from the toolbar, or pull down the "Insert" menu and pick **Sheet Metal – Base Flange**.

Select the **Top** plane when prompted to select a plane.

Create a rectangle with the lower left corner at the origin using the **Rectangle** icon in the CommandManager, or pull down the "Tools" menu and pick **Sketch Entities – Rectangle**.

Add the dimensions as shown using the **Smart Dimension** icon in the CommandManager, or pull down the "Tools" menu and pick **Dimensions – Smart**.

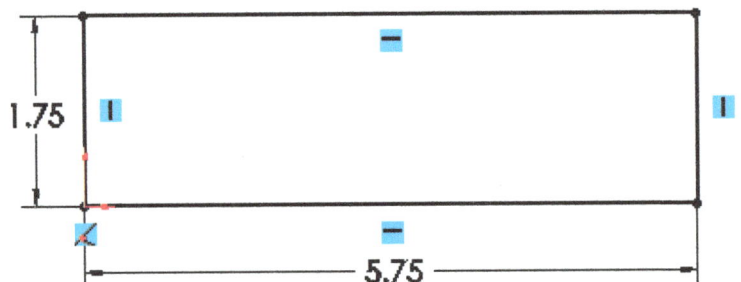

1.75

5.75

Exit the sketch by clicking the **Exit Sketch** icon in the CommandManager or in the upper right corner of the graphics area.

In the **Base Flange** PropertyManager, under **Sheet Metal Gauges**, check the **Use gauge table** check box.

Pick **CRS GAUGES** from the **Select Table** pull down.

Use the default **16 Gauge Thickness**, and make sure that the **Reverse direction** check box is checked.

Under **Bend Allowance**, pick **Bend Table** from the pull down menu. Then pick **SMG KFACTOR BASE BEND TABLE** from the **Bend Table** pull down menu.

Click the green check mark button at the top of the **Base Flange** PropertyManager.

Click the **Sheet Metal** icon in the control area of the CommandManager. Then, click the **Sketched Bend** icon from the toolbar, or pull down the "Insert" menu and pick **Sheet Metal – Sketched Bend**.

Select the top of the part.

In the bottom left corner of the graphics area, change the View orientation by clicking the pull down arrow and picking **Top**, or press **Ctrl+5**.

Create two parallel angled lines using the **Line** icon in the CommandManager, or pull down the "Tools" menu and pick **Sketch Entities – Line**.

Add the dimensions as shown using the **Smart Dimension** icon in the CommandManager, or pull down the "Tools" menu and pick **Dimensions – Smart**.

Exit the sketch by clicking the **Exit Sketch** icon in the CommandManager or in the upper right corner of the graphics area.

Click on the top of the part in between the parallel lines.

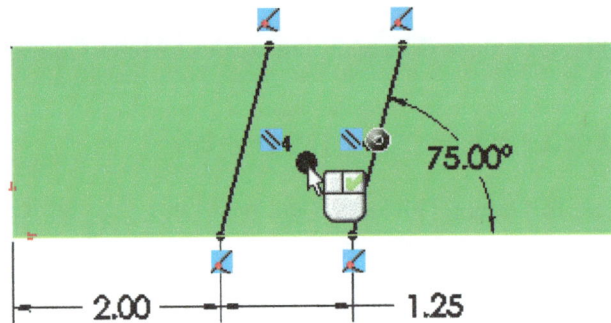

In the **Sketched Bend** PropertyManager, make sure that the **Bend Outside** button is selected.

Enter '**45**' for the **Bend Angle**.

Click the green check mark button at the top of the **Sketched Bend** PropertyManager.

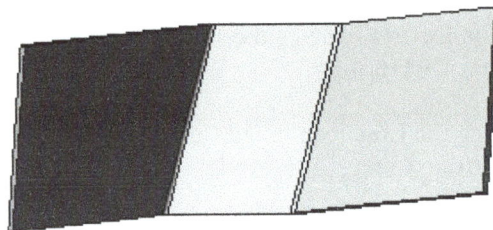

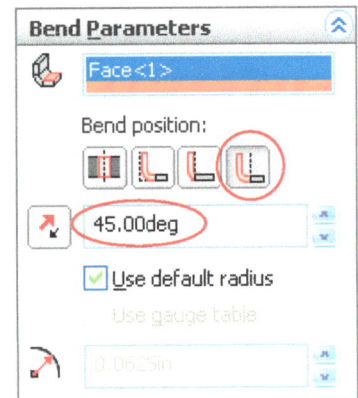

Click the **Flatten** icon in the CommandManager, or right click on **Flat-Pattern1** in the FeatureManager design tree and pick **Unsuppress**.

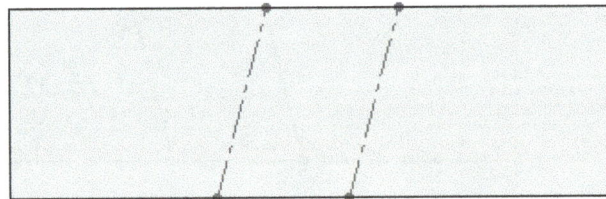

The result here is a straight edge part. This simplifies the shape of the blank and reduces the manufacturing cost.

Saving the Part

Click the **Save** icon in the "Standard" toolbar, or pick **Save** from the "File" pull down menu.

In the **Save As** dialog box, in the **File name** box, type '**Chapter 5 – Sketched Bend**'.

Chapter 6

Unrolling a Cylinder

When the part shape is a bend area with no flanges, you need to use the Unfold and Fold commands to better model the cutouts and features. Unlike Chapter 3, here we want the cutouts to be the true shapes in the flat.

Here you will model a simple hose clamp to learn how this works. Note that the Feature command Fillet is used rather than the Sheet Metal command Break Corner. The Break Corner command only works on formed/folded parts and then only on the flanges, not the bend areas.

Create the Base Flange

Begin a new **Part** document by clicking the **New** icon in the "Standard" toolbar, or pull down the "File" menu and pick **New**.

In the FeatureManager design tree, right click on **Material**, and pick **Plain Carbon Steel** from the menu.

Create a base flange by clicking the **Sheet Metal** icon in the control area of the CommandManager. Then, click the **Base-Flange/Tab** icon from the toolbar, or pull down the "Insert" menu and pick **Sheet Metal – Base Flange**.

Select the **Front** plane when prompted to select a plane on which to sketch the feature cross-section.

Create a **4.00** circle centered on the origin using the **Circle** icon in the CommandManager, or pull down the "Tools" menu and pick **Sketch Entities – Circle**.

Create a centerline line through the origin as shown using the **Centerline** icon in the CommandManager, or pull down the "Tools" menu and pick **Sketch Entities – Centerline**.

Then, create a centerline on each side of the first centerline as shown

Add two **.125** dimensions between the centerlines as shown using the **Smart Dimension** icon in the CommandManager, or pull down the "Tools" menu and pick **Dimensions – Smart**.

Then, click on the **Trim Entities** icon from the CommandManager, or pull down the "Tools" menu and pick **Sketch Tools – Trim**.

With **Trim to closest** selected, click on the two small parts of the circle in between the centerlines to trim them as shown.

Exit the sketch by clicking the **Exit Sketch** icon in the CommandManager or in the upper right corner of the graphics area.

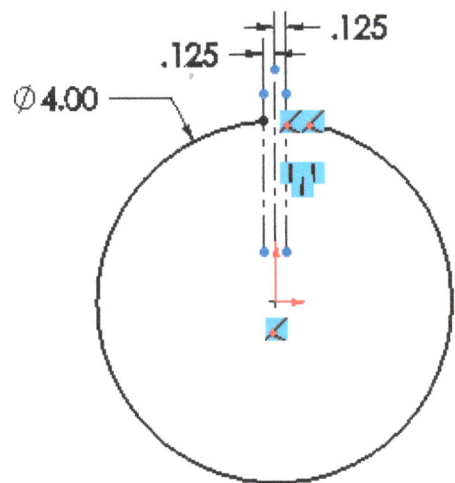

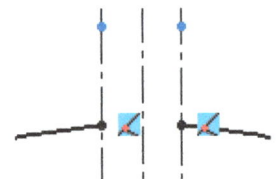

In the **Base Flange** PropertyManager under **Direction 1**, set the **End Condition** to **Blind** and the **Depth** to '**.5**'.

Under **Sheet Metal Gauges**, check the **Use gauge table** check box, and pick **CRS GAUGES** from the **Select Table** pull down.

Under **Sheet Metal Parameters**, pick **18 Gauge** for the **Thickness**, and make sure that the **Reverse direction** check box is not checked.

Under **Bend Allowance**, make sure that **K-Factor** is selected. The default value of '**0.3333**' should appear in the window below this. If not, check the **Override value** box and set the value to '**0.3333**'.

✓ Click the green check mark button at the top of the **Base Flange** PropertyManager to accept the settings and create the part.

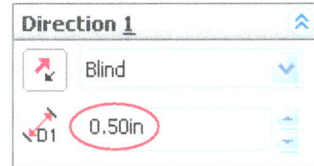

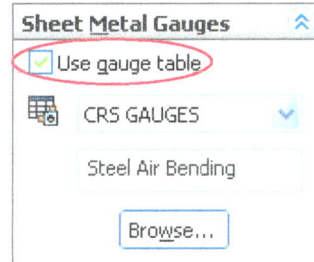

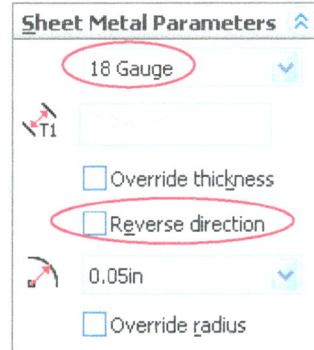

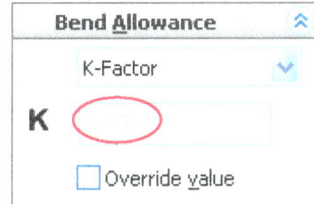

Unfold the Clamp

To unfold the clamp, click the **Unfold** icon in the CommandManager, or pull down the "Insert" menu and pick **Sheet Metal – Unfold**.

For the **Fixed face**, select the edge of the cutout as shown.

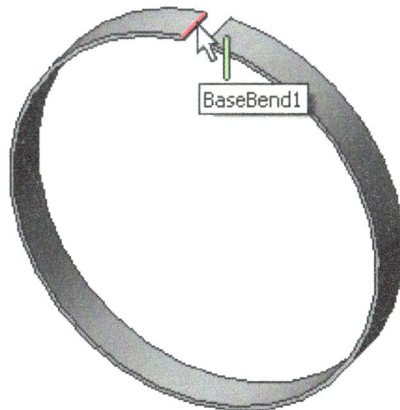

For the **Bends to unfold**, click the **Collect All Bends** button.

✓ Click the green check mark button at the top of the **Unfold** PropertyManager and the bends will unfold.

In the bottom left corner of the graphics area, change the View orientation by clicking the pull down arrow and picking **Top**, or press **Ctrl+5**.

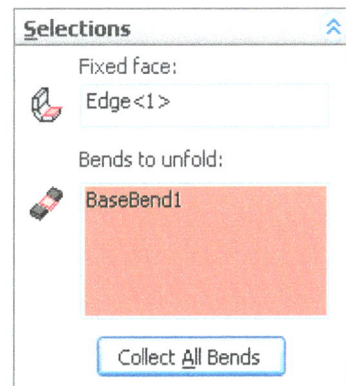

Chamfer the Corners

Change back to the **Trimetric** view by clicking the View orientation pull down arrow in the bottom left corner of your graphics area and picking **Trimetric**.

Use the **Zoom to Area** command to zoom in on the right end of the part.

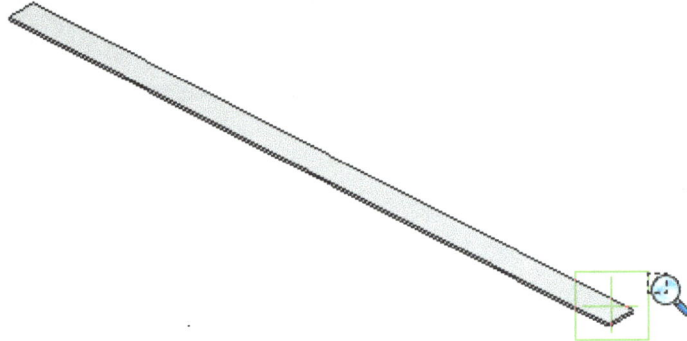

Click the **Features** icon in the control area of the CommandManager. Then, click the **Chamfer** icon from the toolbar, or pull down the "Insert" menu and pick **Features – Chamfer**.

Select the two vertical material thickness edge lines at the right end of the part.

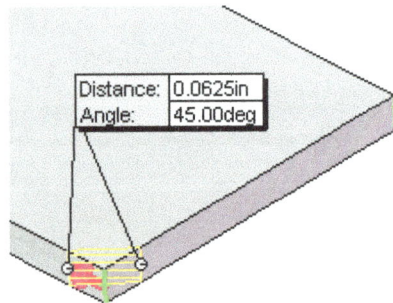

Set the **Distance** to '**0.0625**' and make certain the **Angle** value is set to **45** degrees.

Click the green check mark button at the top of the **Chamfer** PropertyManager.

In the bottom left corner of the graphics area, change the View orientation by clicking the pull down arrow and picking **Top**, or press **Ctrl+5**.

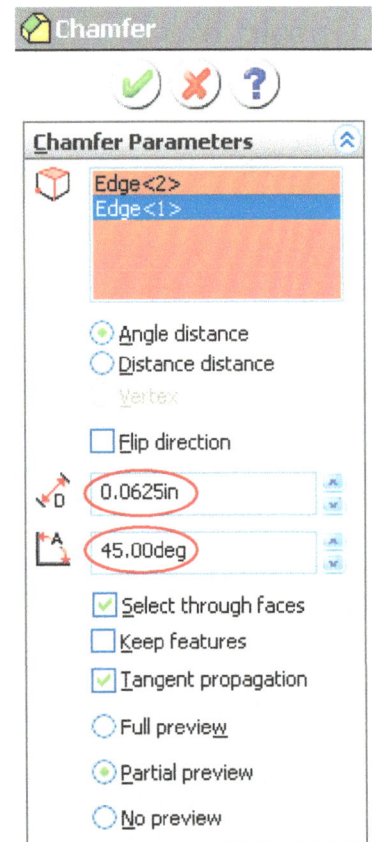

Create a Cutout Pattern

Click the **Extruded Cut** icon from the CommandManager or pull down the "Insert" menu and pick **Cut – Extrude**.

Select the top of the part.

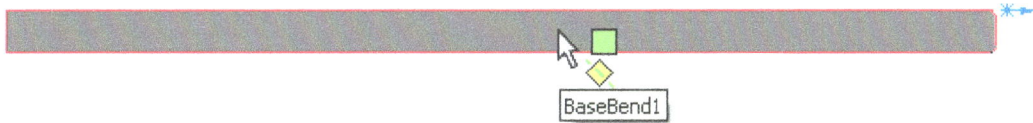

Create a horizontal centerline line through the midpoint of the right vertical edge using the **Centerline** icon in the CommandManager, or pull down the "Tools" menu and pick **Sketch Entities – Centerline**.

Create two vertical lines on the far right of the part as shown using the **Line** icon in the CommandManager, or pull down the "Tools" menu and pick **Sketch Entities – Line**.

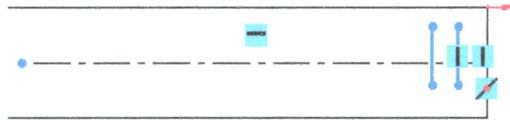

Press the **Escape** key and then hold down the **Ctrl** key and select both of these lines. In the **Properties** PropertyManager, click on the **Equal** button.

Right click on the right vertical line and pick **Select Midpoint** from the menu. Then, hold down the **Ctrl** key and select the centerline.

In the **Properties** PropertyManager, click on the **Coincident** button.

Repeat this for the left vertical line.

Click on the **Tangent Arc** icon in the CommandManager, or pull down the "Tools" menu and pick **Sketch Entities – Tangent Arc**.

Connect the top and bottom of the vertical lines using tangent arcs. You may need to zoom in closer to see the endpoints needed to create the arcs.

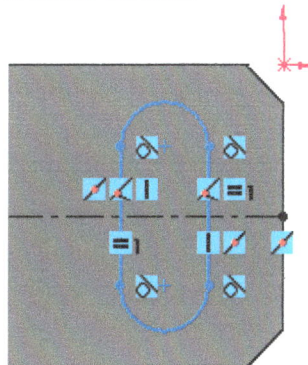

Click on the **Smart Dimension** icon in the CommandManager, or pull down the "Tools" menu and pick **Dimensions – Smart**.

Add a '**.03125**' dimension to the top arc. Then, add a '**.125**' dimension between the right vertical line and the right of the part. Next, add a '**.375**' dimension between the two arcs. You must pick the two arcs here and not just dimension the vertical line.

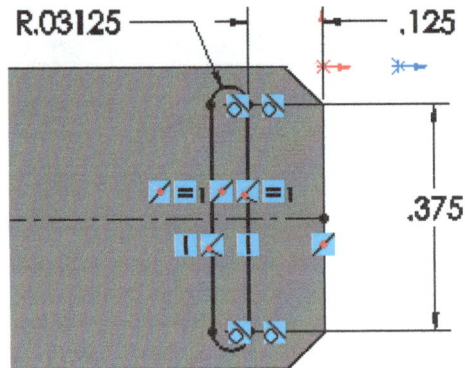

Press the **Escape** key, and then, right click on the **.375** dimension and pick **Properties** from the menu.

In the bottom right of the **Dimension Properties** dialog box, change the **First arc condition** and the **Second arc condition** to **Max**.

Then, click the **OK** button.

Double click on the new **.4375** dimension and change it back to '**.375**'.

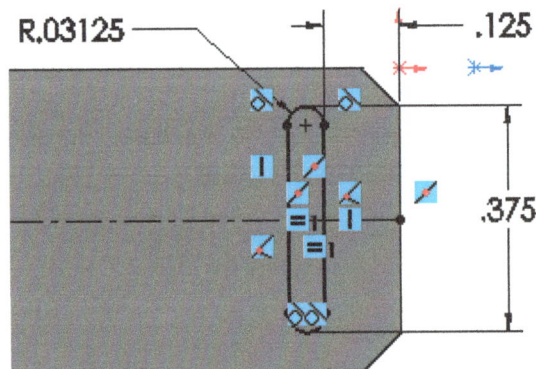

Pull down the "Tools" menu and pick **Sketch Tools – Linear Pattern**.

Select the two lines and two arcs.

In the **Linear Pattern** PropertyManager, under **Direction 1**, click on the **Reverse Direction** icon and set the **Spacing** to '**.125**'.

Then, set the **Number** to '**25**'.

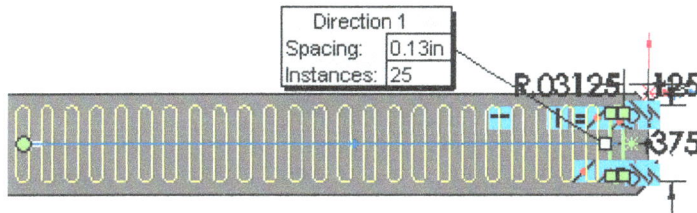

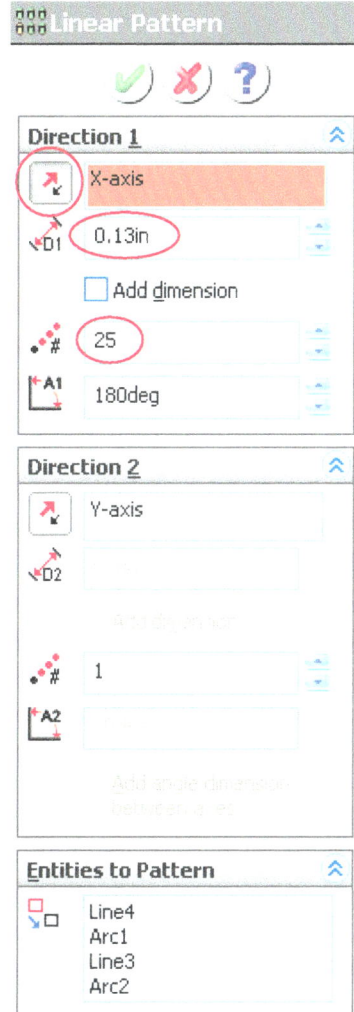

Direction 1	
Spacing:	0.13in
Instances:	25

R.03125 .125

.375

Click the green check mark button at the top of the **Linear Pattern** PropertyManager.

Exit the sketch by clicking the **Exit Sketch** icon in the CommandManager or in the upper right corner of the graphics area.

In the **Cut-Extrude** PropertyManager, check the **Link to thickness** check box, and then, click the green check mark button at the top of the **Cut-Extrude** PropertyManager.

Linear Pattern

Direction 1

X-axis

D1 0.13in

☐ Add dimension

25

A1 180deg

Direction 2

Y-axis

D2

1

A2

Add angle dimension between a #5

Entities to Pattern

Line4
Arc1
Line3
Arc2

Create Another Cutout

Click the **Extruded Cut** icon from the CommandManager, or pull down the "Insert" menu and pick **Cut – Extrude**.

Select the top of the part.

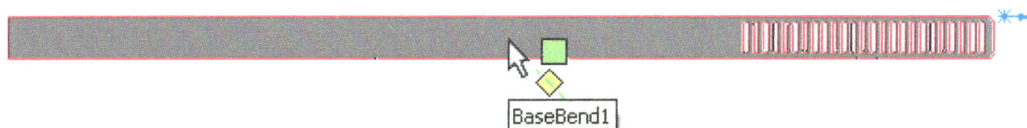

BaseBend1

Use the **Zoom to area** command to zoom in on the left end of the part.

Create a horizontal centerline line through the midpoint of the left vertical edge using the **Centerline** icon in the CommandManager, or pull down the "Tools" menu and pick **Sketch Entities – Centerline**.

Create the sketch as shown below using the **Line** icon and the **Tangent Arc** icon in the CommandManager, or pull down the "Tools" menu and pick **Sketch Entities – Line** and **Sketch Entities – Tangent Arc**.

Ctrl select the center point of the right arc and the horizontal centerline.

In the **Properties** PropertyManager, click the **Coincident** button.

Then, select the other arc centerpoint. Hold down the **Ctrl** key and select the centerline.

In the **Properties** PropertyManager, click the **Coincident** button.

Ctrl select the two vertical lines. In the **Properties** PropertyManager, click the **Collinear** button.

Next, add the dimensions as shown below.

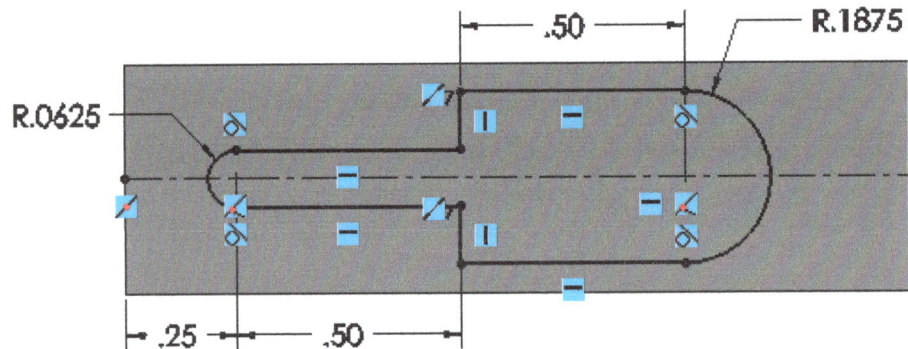

Pull down the "View" menu and pick **Sketch Relations** to toggle the display of the sketch relation flags.

Then, add '.03' fillets to the inside corners using the **Sketch Fillet** icon in the CommandManager, or pull down the "Tools" menu and pick **Sketch Tools – Fillet**.

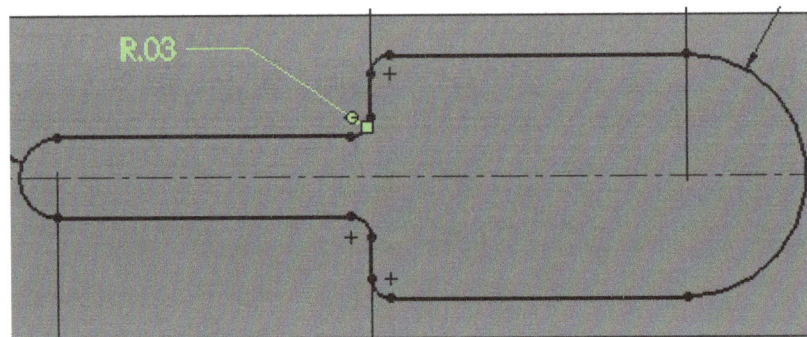

Pull down the "View" menu and pick **Sketch Relations** to toggle the display of the sketch relation flags.

Exit the sketch by clicking the **Exit Sketch** icon in the CommandManager or in the upper right corner of the graphics area.

In the **Cut-Extrude** PropertyManager, check the **Link to thickness** check box, and then, click the green check mark button at the top of the **Cut-Extrude** PropertyManager.

Click the **Sheet Metal** icon in the control area of the CommandManager. Then, click the **Fold** icon in the CommandManager, or pull down the "Insert" menu and pick **Sheet Metal – Fold**.

In the **Fold** PropertyManager, click on the **Collect All Bends** button to select all the appropriate bends in the part.

Click the green check mark button at the top of the **Fold** PropertyManager to fold the part back up.

Change back to the **Trimetric** view by clicking the View orientation pull down arrow in the bottom left corner of your graphics area and picking **Trimetric**.

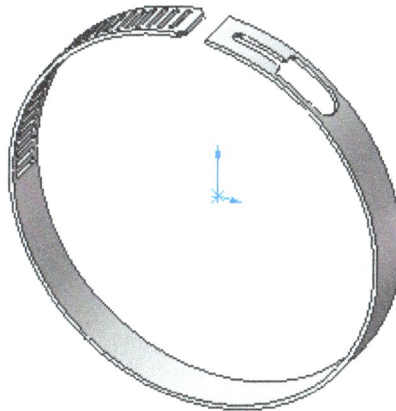

Click the **Flatten** icon in the CommandManager, or right click on **Flat-Pattern1** in the FeatureManager design tree and pick **Unsuppress**.

In the bottom left corner of the graphics area, change the View orientation by clicking the pull down arrow and picking **Top**, or press **Ctrl+5**.

Click the **Flatten** icon in the CommandManager, right click on **Flat-Pattern1** in the FeatureManager design tree and pick **Suppress**.

Change back to the **Trimetric** view by clicking the View orientation pull down arrow in the bottom left corner of your graphics area and picking **Trimetric**.

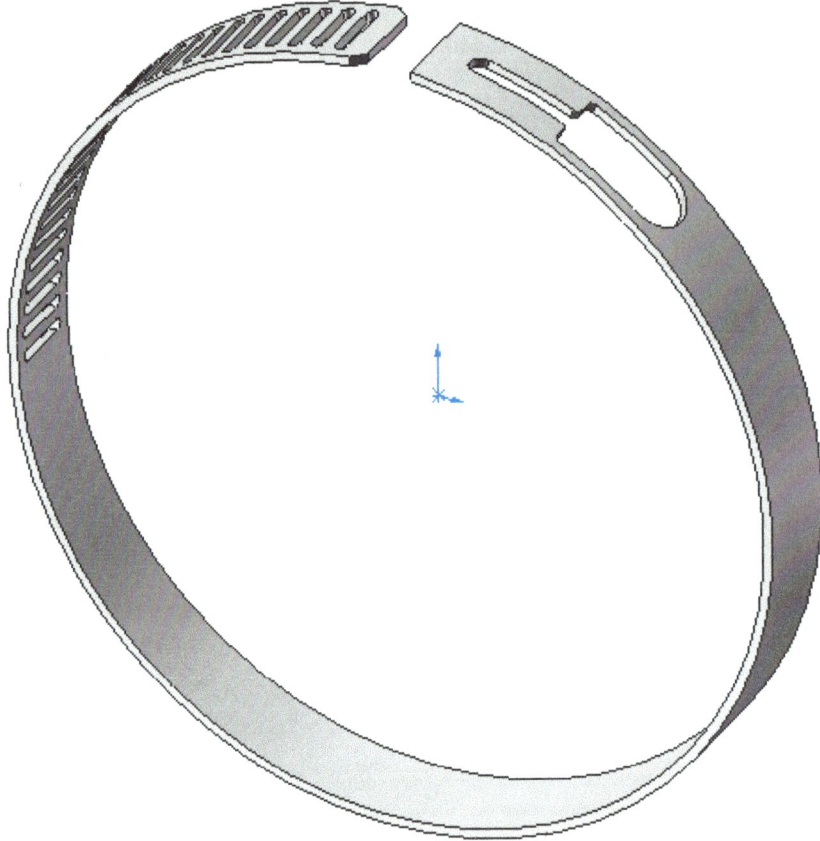

Saving the Part

Click the **Save** icon in the "Standard" toolbar, or pick **Save** from the "File" pull down menu.

In the **Save As** dialog box, in the **File name** box, type '**Hose Clamp**' and click **Save**.

Chapter 7

Conical Faces

Sheet metal cones are a common part for many users. SolidWorks says as long as the part is a true cone you can do it.

This sample starts with a standard cone. Then, you will cut an angle across the top. Once you have created one, all you need to do is change the dimensions and rebuild to make all the rest.

A thicker material is used here just to show you the thickness is not important. It is the shape and method that matter the most. And when you get to the gap, or seam, don't over do it. The 359.9 sweep of this cone leaves a gap distance of 0.01. Not bad, considering the tolerance of your manufacturing processes.

Create a Sheet Metal Part with Conical Faces

⬜ Begin a new **Part** document by clicking the **New** icon in the "Standard" toolbar, or pull down the "File" menu and pick **New**.

In the FeatureManager design tree, right click on **Material**, and pick **Plain Carbon Steel** from the menu.

Click the **Features** icon in the control area of the CommandManager. Then, click the **Revolved Boss/Base** icon from the toolbar, or pull down the "Insert" menu and pick **Boss/Base – Revolve**.

Select the **Front** plane when prompted to select a plane on which to sketch the feature cross-section.

Create a vertical centerline line through the origin using the **Centerline** icon in the CommandManager, or pull down the "Tools" menu and pick **Sketch Entities – Centerline**.

Create an angled line to the right of the origin using the **Line** icon in the CommandManager, or pull down the "Tools" menu and pick **Sketch Entities – Line**.

Press the **Escape** key to deselect the line.

Then, **Ctrl** select the origin and the bottom right endpoint of the line. In the **Properties** PropertyManager, click the **Horizontal** button.

Click the **Smart Dimension** icon in the CommandManager, or pull down the "Tools" menu and pick **Dimensions – Smart**.

Add a '5' and '8' horizontal dimension between the centerline and the end of the angled line. Make sure that you pick the centerline and not the endpoint of the centerline. This will ensure that a radial dimension will be placed instead of a linear dimension. Then, add a '**14**' vertical dimension between the two endpoints of the angled line.

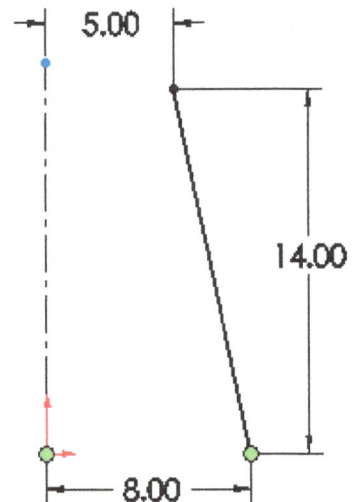

Once the dimensions are placed, press **Escape**, and then, right click on the **5.00** dimension and pick **Properties** from the menu.

In the **Dimension Properties** dialog box, check **Diameter Dimension** and click the **OK** button.

Then, right click on the **8.00** dimension and pick **Properties** from the menu.

In the **Dimension Properties** dialog box, check **Diameter Dimension** and click the **OK** button.

☐ Display with parentheses
☐ Display as dual dimension
☐ Display as inspection dimension
☐ Read only
☐ Driven
☑ Diameter dimension

Exit the sketch by clicking the **Exit Sketch** icon in the CommandManager or in the upper right corner of the graphics area.

In the **SolidWorks** dialog box that asks to automatically close the sketch, click the **No** button.

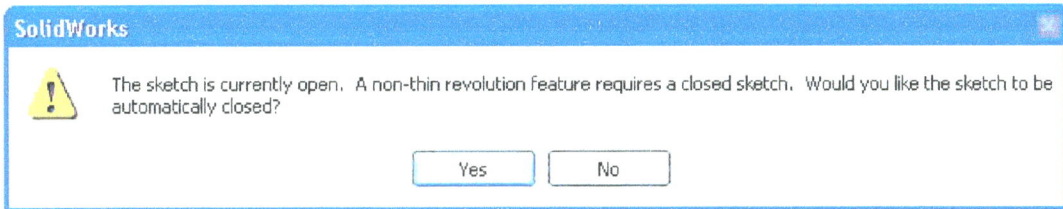

SolidWorks

⚠ The sketch is currently open. A non-thin revolution feature requires a closed sketch. Would you like the sketch to be automatically closed?

[Yes]   [No]

In the **Revolve** PropertyManager, for the **Axis of Revolution**, make sure that the centerline is selected. A preview of the revolve will appear if the centerline is selected.

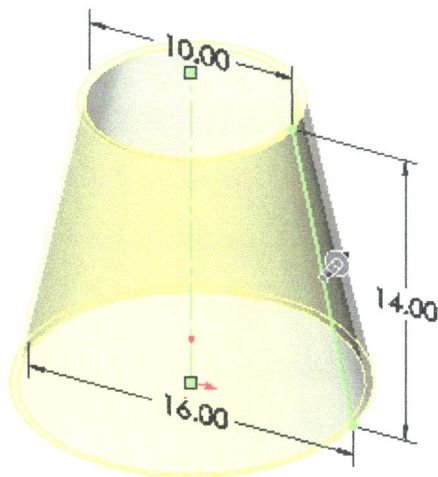

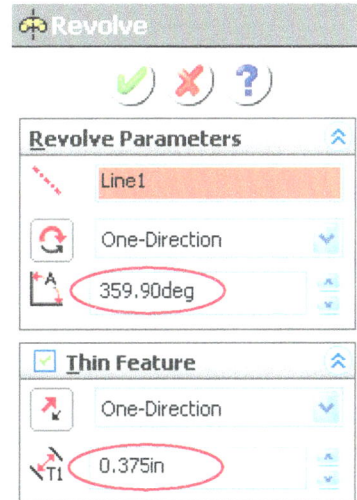

⚙ Revolve

✔ ✖ ?

Revolve Parameters

Line1

One-Direction

359.90deg

☐ Thin Feature

One-Direction

0.375in

Make sure that the **Revolve Type** is set to **One-Direction**, and enter '**359.9**' for the **Angle**.

Under **Thin Feature**, make sure that the **Type** is set to **One-Direction**, and enter '**.375**' for the **Direction 1 Thickness**.

Click the green check mark button at the top of the **Revolve** PropertyManager to accept the settings and create the part.

Click the **Sheet Metal** icon in the control area of the CommandManager. Then, click the **Insert Bends** icon from the toolbar, or pull down the "Insert" menu and pick **Sheet Metal - Bends**.

In the **Bends** PropertyManager, under **Bend Parameters**, make sure that the **Fixed Face or Edge** box is selected.

Then, in the graphics area, select the seam edge line as shown. The fixed edge remains in place when the part is flattened.

Since you are creating a sheet metal part with one or more conical faces, you must select **K-Factor** as the type of **Bend Allowance**. Enter '**.5**' for the **K-Factor**.

These options and values that you specify will be shown as the default settings for the next new sheet metal part that you create.

Click the green check mark button at the top of the **Bends** PropertyManager.

Click the **Flatten** icon in the CommandManager, or right click on **Process-Bends1** and pick **Suppress**.

Select the top of the part. Then, in the bottom left corner of the graphics area, change the View orientation by clicking the pull down arrow and picking **Normal To**, or press **Ctrl+8**.

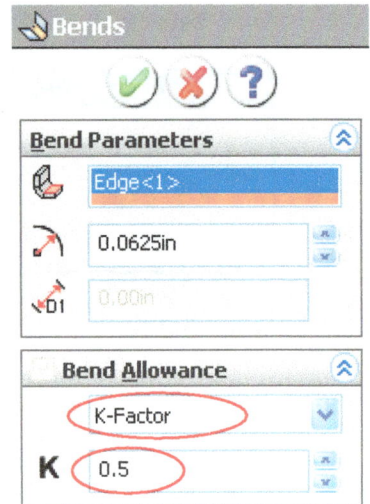

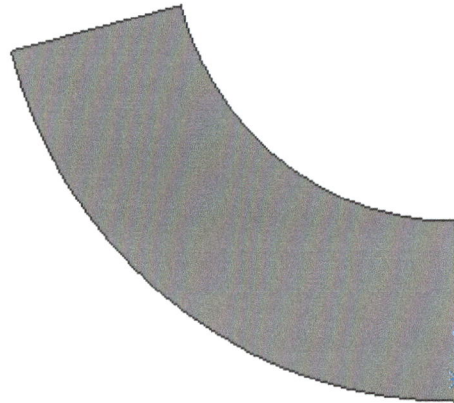

Click the **Flatten** icon in the CommandManager, or right click on **Process-Bends1** and pick **Unsuppress**.

Add a Cut Extrude

Now let's trim the top of the cone, making it a little irregular. Click the **Extruded Cut** icon from the CommandManager or pull down the "Insert" menu and pick **Cut – Extrude**.

In the upper left corner of the graphics area, click the plus sign next to **Part1**.

Select **Front** plane from the flyout FeatureManager design tree.

In the bottom left corner of the graphics area, change the View orientation by clicking the pull down arrow and picking **Front**, or press **Ctrl+1**.

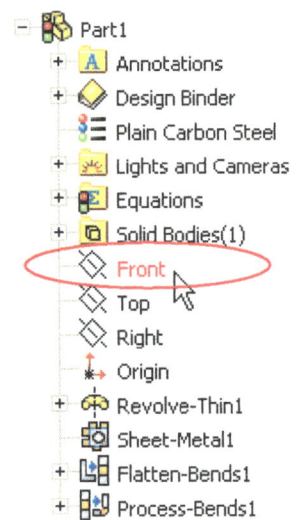

Click the **Line** icon in the CommandManager, or pull down the "Tools" menu and pick **Sketch Entities – Line**.

Create an angled line beginning at the upper left corner of the part as shown. To collapse the flyout FeatureManager design tree, click on the negative sign next to **Part1**. Make sure the line goes all the way across the part as shown.

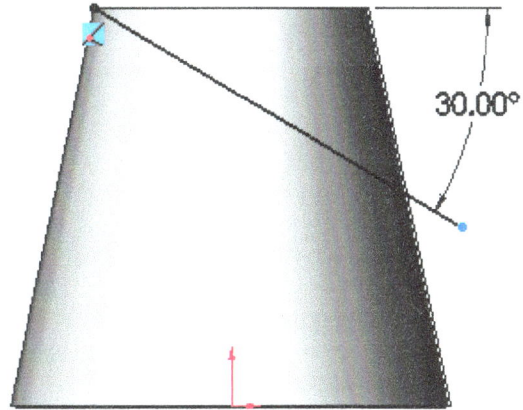

Add a '**30**' angle dimension between the line and the top of the part using the **Smart Dimension** icon in the CommandManager, or pull down the "Tools" menu and pick **Dimensions – Smart**.

Exit the sketch by clicking the **Exit Sketch** icon in the CommandManager or in the upper right corner of the graphics area.

In the **Cut-Extrude** dialog box, set the **End Condition** of **Direction 1** and **Direction 2** to **Through All**.

Click the green check mark button at the top of the **Cut-Extrude** PropertyManager.

Zoom in to the top edge of the part. Note that the edge of the part is not perpendicular to the outer surface of the cone. This is due to the Cut-Extrude being done after the sheet metal feature was processed.

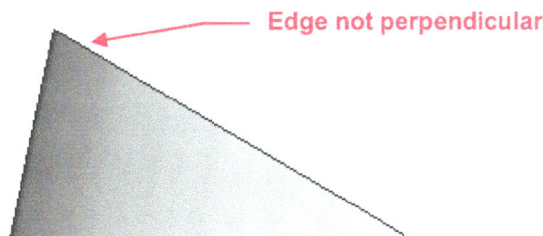

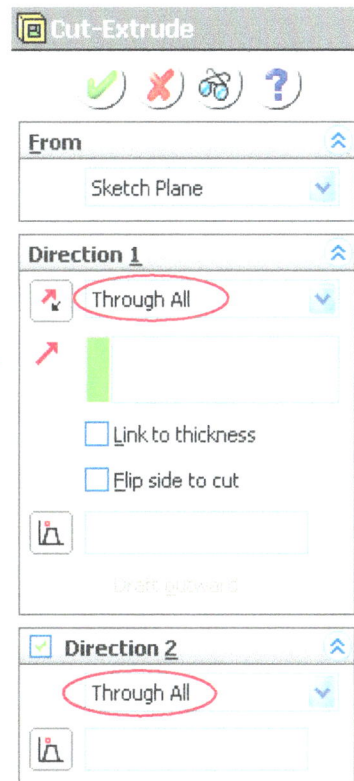

Edge not perpendicular

In the bottom left corner of the graphics area, change the View orientation by clicking the pull down arrow and picking **Trimetric**.

Click the **Flatten** icon in the CommandManager, or right click on **Process-Bends1** and pick **Suppress**.

Select the top of the part. Then, in the bottom left corner of the graphics area, change the View orientation by clicking the pull down arrow and picking **Normal To**, or press **Ctrl+8**.

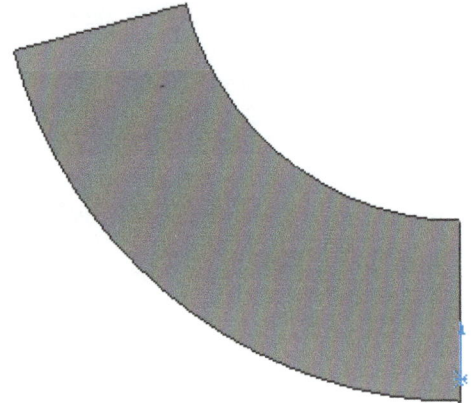

The flat pattern has not changed. This is incorrect. The Cut-Extrude did not alter the flat pattern because, again, the Cut-Extrude was done after the sheet metal feature was processed.

Click the **Flatten** icon in the CommandManager, or right click on **Process-Bends1** and pick **Unsuppress**.

In the FeatureManager design tree, right click on **Cut-Extrude1** and pick **Suppress**.

Right click on **Sheet-Metal1** and pick **Rollback** to roll the model back to an earlier state, suppressing the features below the rollback bar. You may also drag the rollback bar, the yellow and black line which turns blue when selected, to the desired position in the FeatureManager design tree.

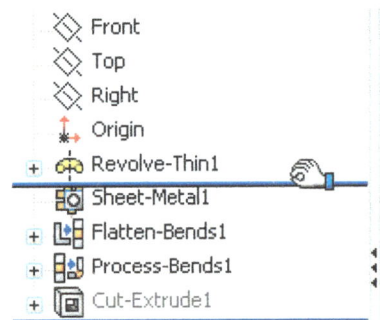

In the bottom left corner of the graphics area, change the View orientation by clicking the pull down arrow and picking **Front**, or press **Ctrl+1**.

Add Another Cut Extrude

If you trim the cone before inserting the bends, SolidWorks will properly define the sheet metal edges and the flat pattern.

Click the **Extruded Cut** icon from the CommandManager or pull down the "Insert" menu and pick **Cut – Extrude**.

In the upper left corner of the graphics area, click the plus sign next to **Part1**.

Select **Front** plane from the flyout FeatureManager design tree.

Create an angled line beginning at the upper left corner of the part as shown using the **Line** icon in the CommandManager, or pull down the "Tools" menu and pick **Sketch Entities – Line**. To collapse the flyout FeatureManager design tree, click on the negative sign next to **Part1**. Make sure the line goes all the way across the part as shown.

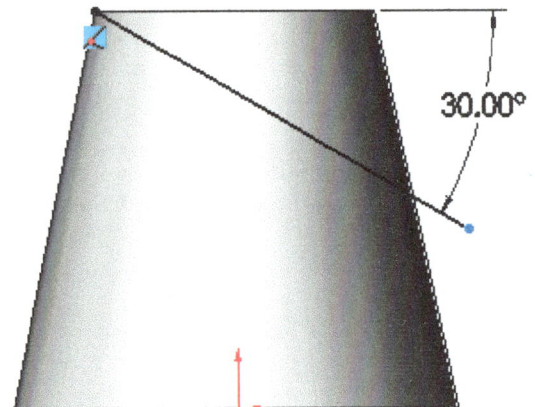

Add a '30' angle dimension between the line and the top of the part using the **Smart Dimension** icon in the CommandManager, or pull down the "Tools" menu and pick **Dimensions – Smart**.

Exit the sketch by clicking the **Exit Sketch** icon in the CommandManager or in the upper right corner of the graphics area.

In the **Cut-Extrude** dialog box, set the **End Condition** of **Direction 1** and **Direction 2** to **Through All**, as you did before.

Click the green check mark button at the top of the **Cut-Extrude** PropertyManager.

In the FeatureManager design tree, right click on **Cut-Extrude2** and pick **Roll to End**.

A warning dialog appears. Click **Stop and Repair**.

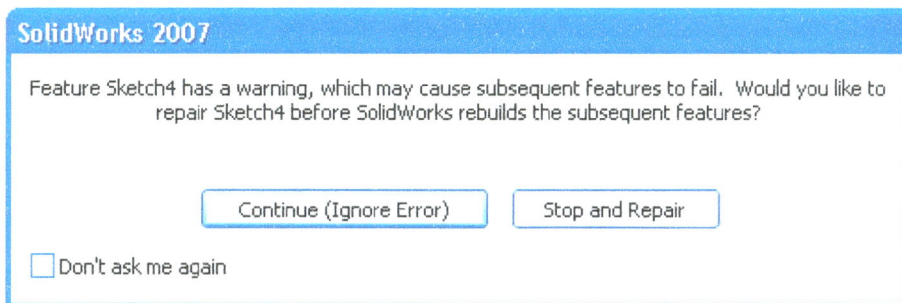

SolidWorks 2007

Feature Sketch4 has a warning, which may cause subsequent features to fail. Would you like to repair Sketch4 before SolidWorks rebuilds the subsequent features?

Continue (Ignore Error)   Stop and Repair

☐ Don't ask me again

The problem is with **Sketch4**. **Cut-Extrude1** is suppressed, but the sketch used to create the feature is not. Because the new Cut-Extrude modified the part used to create **Sketch4**, an error occurs.

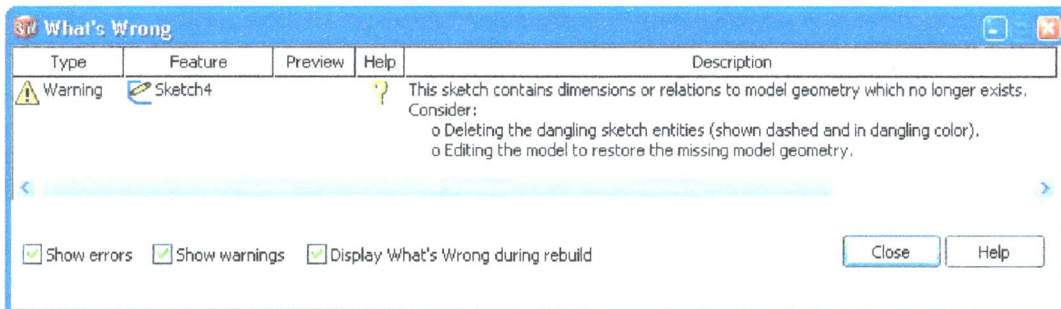

Type	Feature	Preview	Help	Description
⚠ Warning	Sketch4		?	This sketch contains dimensions or relations to model geometry which no longer exists. Consider: o Deleting the dangling sketch entities (shown dashed and in dangling color). o Editing the model to restore the missing model geometry.

☑ Show errors ☑ Show warnings ☑ Display What's Wrong during rebuild   Close   Help

You will not fix this error here, but rather eliminate it by suppressing the sketch. Click the plus sign next to **Cut-Extrude1** in the FeatureManager design tree to expand the feature.

Right click on **Sketch4** and pick **Suppress**. **Sketch4** is suppressed and the **What's Wrong** dialog box, clearing the error.

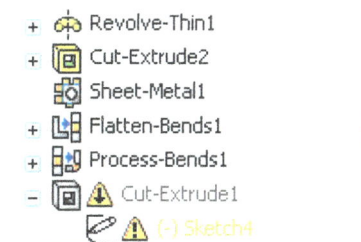

+ Revolve-Thin1
+ Cut-Extrude2
  Sheet-Metal1
+ Flatten-Bends1
+ Process-Bends1
– Cut-Extrude1
    (-) Sketch4

Zoom in to the top edge of the part. Note the edge is now perpendicular to the outside surface of the cone. This is due to the Cut-Extrude being done before the sheet metal feature was processed.

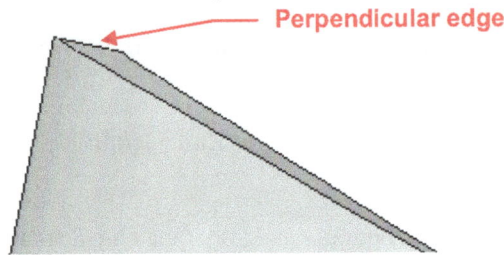

Perpendicular edge

Let's take a look directly at this new top edge of the cone.

In the bottom left corner of the graphics area, change the View orientation by clicking the pull down arrow and picking **Right**, or press **Ctrl+4**.

Click the **Sketch** icon in the control area of the CommandManager. Then, click the **Sketch** icon from the toolbar, or pull down the "Insert" menu and pick **Sketch**.

In the upper left corner of the graphics area, click the plus sign next to **Part1**.

Select **Right** plane from the flyout FeatureManager design tree.

Click the **Line** icon in the CommandManager, or pull down the "Tools" menu and pick **Sketch Entities – Line**.

Select the right midpoint of the outside circular edge as the starting point of the line as shown. Make certain the red dot is visible and not the red circle around the cone.

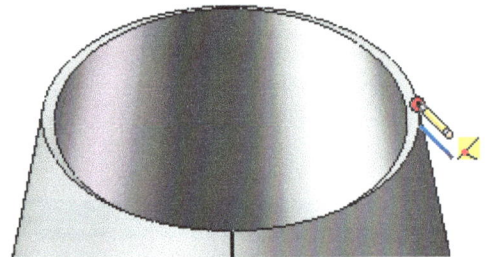

Then, select the left midpoint of the circular edge as the ending point of the line as shown.

Exit the sketch by clicking the **Exit Sketch** icon in the CommandManager or in the upper right corner of the graphics area.

Pull down the "Insert" menu and pick **Reference Geometry – Plane**.

Select the line and the bottom end point of the top edge of the cone as shown.

Click the green check mark button at the top of the **Plane** PropertyManager.

Then, in the bottom left corner of the graphics area, change the View orientation by clicking the pull down arrow and picking **Normal To**.

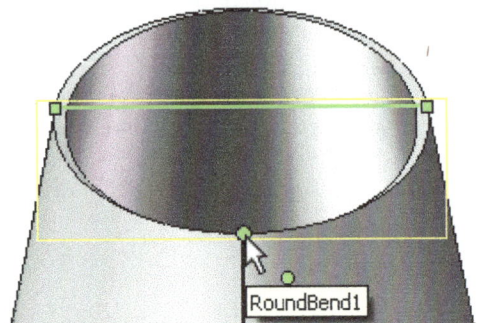

As you can see, the top of the cone is not a circle, but it is actually an ellipse.

Click the **Sheet Metal** icon in the control area of the CommandManager. Then, click the **Flatten** icon from the toolbar. Normally you could also right click on **Process-Bends1** and pick **Suppress**. However, this will result in an error here since the new sketch is dependant on **Process-Bends1** for its definition.

Select the top of the part. Then, in the bottom left corner of the graphics area, change the View orientation by clicking the pull down arrow and picking **Normal To**, or press **Ctrl+8**.

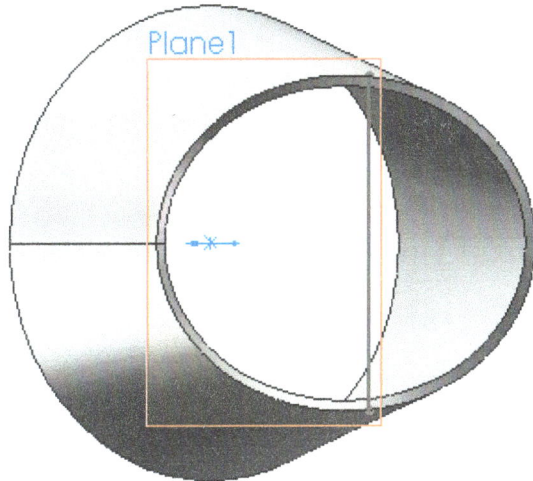

The flat pattern is now correct. **Cut-Extrude2** altered the flat pattern because the Cut-Extrude was done before the sheet metal feature was processed.

Click the **Flatten** icon in the CommandManager.

In the bottom left corner of the graphics area, change the View orientation by clicking the pull down arrow and picking **Trimetric**.

In the FeatureManager design tree, right click on **Sketch6** and pick **Hide**. Then, right click on **Plane1** and pick **Hide**.

Saving the Part

Click the **Save** icon in the "Standard" toolbar, or pick **Save** from the "File" pull down menu.

In the **Save As** dialog box, in the **File name** box, type '**Chapter 7 - Cone**' and click **Save**.

Edit the Dimensions

Now that you have created the sheet metal part, display the dimensions.

To do this, in the FeatureManager design tree, right click on **Annotations** and pick **Show Feature Dimensions**. You may drag and drop dimensions with the mouse to reposition them and make them easier to see. Note that there are two **30.00°** dimensions. One is green because it is from a suppressed sketch.

You may not be able to see all of the dimensions in this view. Rotate the part as needed to access the different dimensions.

To change the size of the cone, double click on the **Ø 10.00** dimension. In the **Modify** dialog box, change the value to '**15**' and press **Enter** or click the green check mark button.

Change the **Ø 16.00** dimension to '**20**', the **14.00** dimension to '**18**', and the black **30.00°** dimension to '**10**'.

To update the model to reflect the changes, click the **Rebuild** icon in the "Standard" toolbar, or press **Ctrl+B**.

You may change the display of the diameter dimensions by right clicking on the dimension. From the menu pick **Display Options – Display As Linear**.

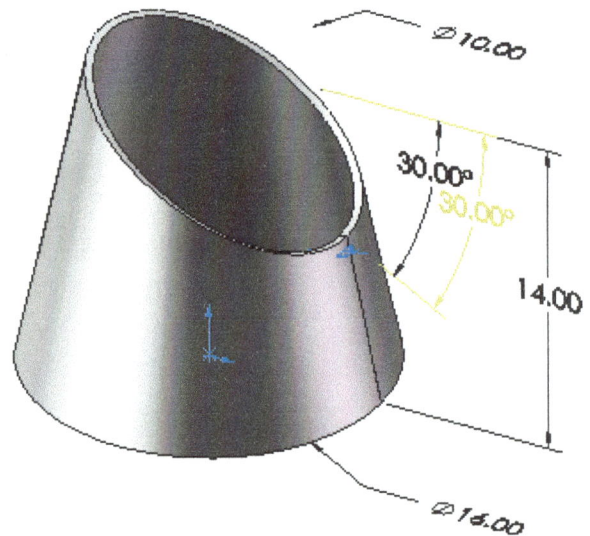

If you get an error rebuilding the model, there are two things to check. First, did you lock the left end of the angled (trim) line onto the top edge of the cone? Second, is the length of the angled line longer than the width of the cone after you changed the dimensions? Both of these can be corrected after altering the sketch of **Cut-Extrude2**. Then, **Rebuild** the model again.

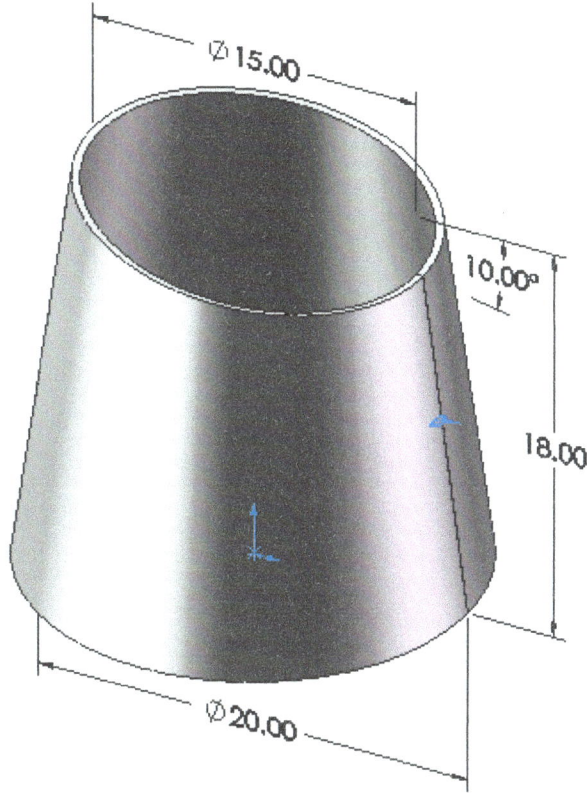

To view the new flat pattern, click the **Sheet Metal** icon in the control area of the CommandManager. Then, click the **Flatten** icon from the toolbar.

Select the top of the part. Then, in the bottom left corner of the graphics area, change the View orientation by clicking the pull down arrow and picking **Normal To**, or press **Ctrl+8**.

You now have the basis to create many different cones. But, remember the angled top edge is not a circle as you may want your cones to be.

Closing the File

Pull down "File" menu and pick **Close**.

Click **No** when prompted to **Save changes to Chapter 7 - Cone.SLDPRT?**.

Chapter 8

Rip Command

Alright, we have avoided it enough. Here comes the Rip command. Although I don't believe a real sheet metal person would ever do this, many users rely heavily on this command.

Create as much of the part as you wish using the solids features. Then, go to the Thin Wall feature to hollow out the part and specify material thickness. Finally, indicate where the rips or tears, as some of you call them, belong.

Be careful when creating a part in this manner. It is easy to model a part which, when unfolded, the flanges overlap one another in the flat.

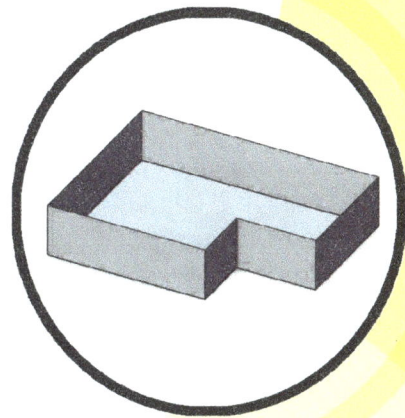

Create the Base

Begin a new **Part** document by clicking the **New** icon in the "Standard" toolbar, or pull down the "File" menu and pick **New**.

Click the **Features** icon in the control area of the CommandManager. Then, click the **Extruded Boss/Base** icon from the toolbar, or pull down the "Insert" menu and pick **Boss/Base – Extrude**.

Select the **Top** plane when prompted to select a plane on which to sketch the feature cross-section.

Create a rectangle with the origin at the lower left corner using the **Rectangle** icon in the CommandManager, or pull down the "Tools" menu and pick **Sketch Entities – Rectangle**.

Create two lines as shown to create a notch in the lower right hand corner of the rectangle using the **Line** icon in the CommandManager, or pull down the "Tools" menu and pick **Sketch Entities – Line**.

Then, click on the **Trim Entities** icon from the CommandManager, or pull down the "Tools" menu and pick **Sketch Tools – Trim**.

With **Trim to closest** selected, click on the two ends of the rectangle to trim the sketch as shown.

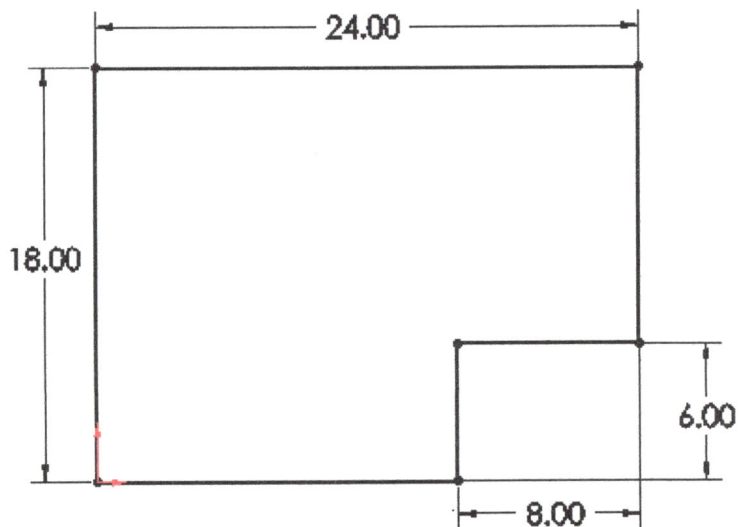

Add the dimensions shown above using the **Smart Dimension** icon in the CommandManager, or pull down the "Tools" menu and pick **Dimensions – Smart**.

Exit the sketch by clicking the **Exit Sketch** icon in the CommandManager or in the upper right corner of the graphics area.

In the **Extrude** PropertyManager, under **Direction 1**, set the **End Condition** to **Blind** and the **Depth** to '5'.

Click the green check mark button at the top of the **Extrude** PropertyManager to accept the settings and create the part.

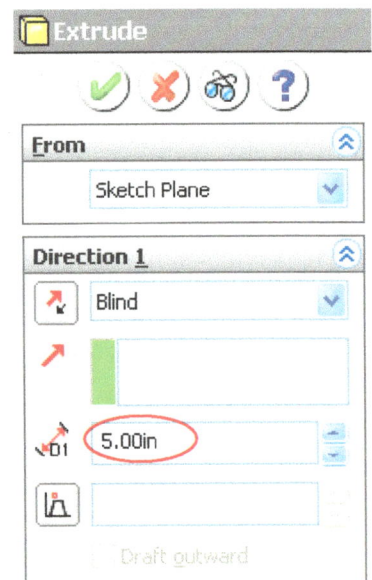

Shell Out the Part

Click the **Shell** icon in the CommandManager, or pull down the "Insert" menu and pick **Features – Shell**.

Select the top of the part.

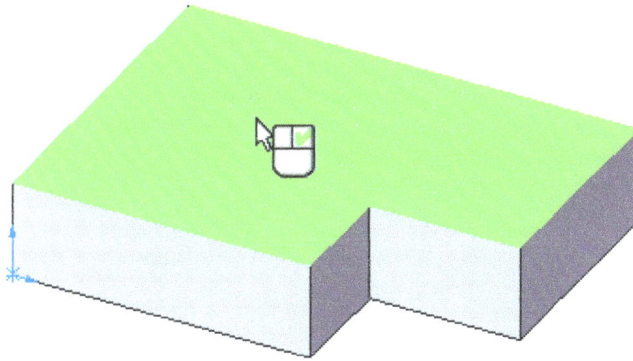

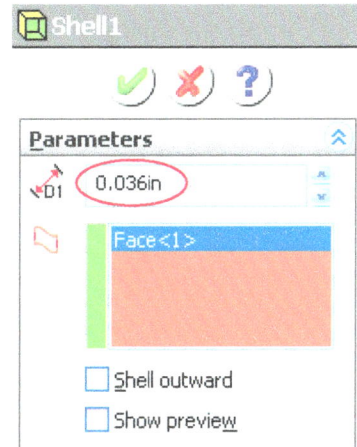

In the **Shell** PropertyManager, under **Parameters**, enter a **Thickness** of '**.036**'.

Click the green check mark button at the top of the **Shell** PropertyManager.

In the FeatureManager design tree, right click on **Material**, and pick **Edit Material** from the menu.

In the **Materials Editor** PropertyManager, under **SolidWorks Materials**, click the plus sign next to **Steel**.

Click on **Chrome Stainless Steel**.

Click the green check mark button at the top of the **Materials Editor** PropertyManager.

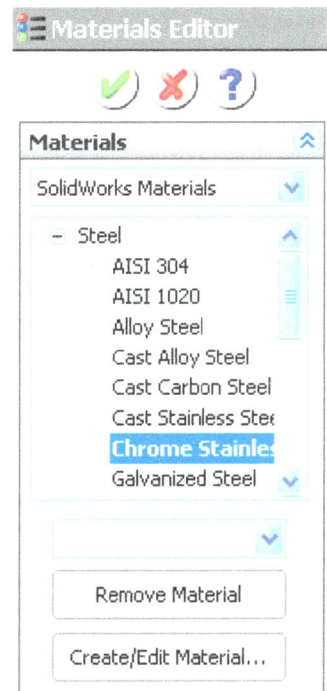

Add a Rip Feature

Click the **Sheet Metal** icon in the control area of the CommandManager. Then, click the **Rip** icon from the toolbar, or pull down the "Insert" menu and pick **Sheet Metal - Rip**.

In the graphics area, click on the edges as shown. To make sure the arrows are pointing in the correct direction, click on the arrow that you want to keep. If the arrow is going the wrong direction, click the arrow to flip its direction. Don't click the arrows if you want to keep both directions.

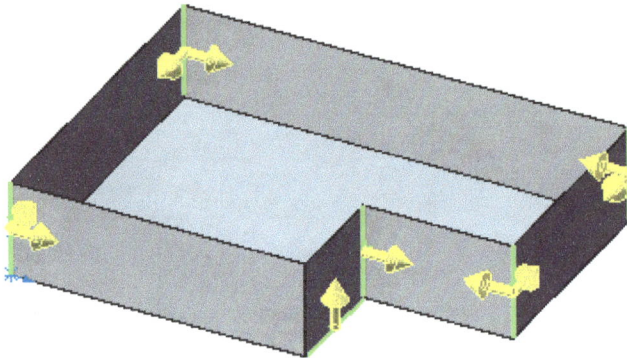

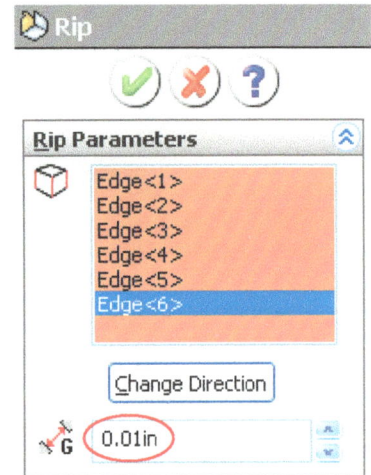

Make sure that the **Rip Gap** is set to '**.01**'.

Click the green check mark button at the top of the **Rip** PropertyManager.

Make the Part Sheet Metal

Click the **Insert Bends** icon in the CommandManager, or pull down the "Insert" menu and pick **Sheet Metal - Bends**.

Select the bottom of the inside of the part.

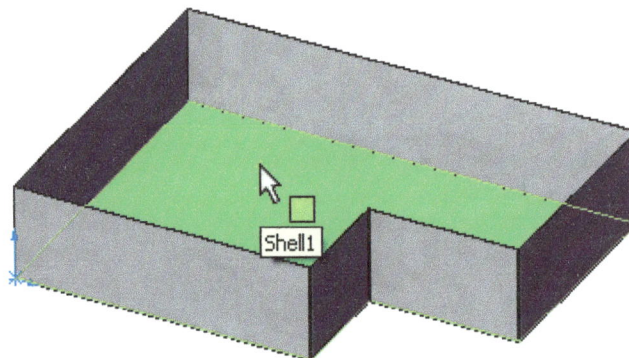

In the **Bends** PropertyManager, under **Bend Parameters**, enter a **Bend Radius** of '**.03125**'.

Under **Bend Allowance**, set the **Bend Allowance Type** to **K-Factor**, and set the **K-Factor** to '**.3333**'.

Make sure **Auto Relief** is checked. Set the **Auto Relief Type** to **Rectangular**, and set the **Relief Ratio** to '**.5**'.

Note that rather than doing a separate **Rip** feature, like you did previously, you can **Insert Bends** and **Rip** in the same command.

Click the green check mark button at the top of the **Bends** PropertyManager.

When the **SolidWorks** dialog box appears, click **OK**.

> **SolidWorks**
>
> ⓘ Auto relief cuts were made for one or more bends.
>
> OK

If it doesn't work, take a close look at the **Rip** edges you picked. One of them is on the bottom of the part, not the side. You can **Undo** this command and try it again if need be.

In the bottom left corner of the graphics area, change the View orientation by clicking the pull down arrow and picking **Top**, or press **Ctrl+5**.

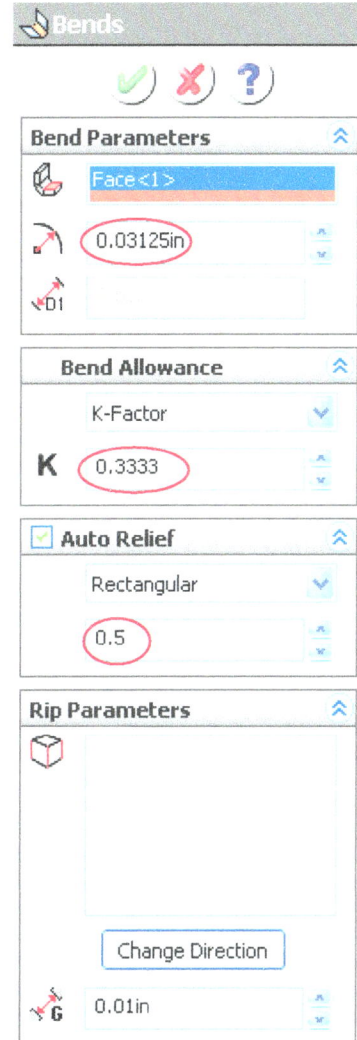

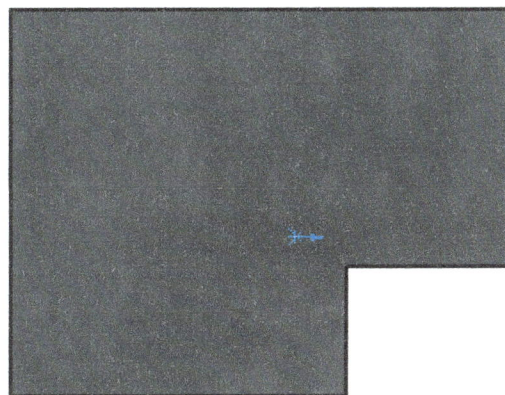

Bends

✓ ✗ ?

Bend Parameters

Face<1>

0.03125in

D1

Bend Allowance

K-Factor

K   0.3333

☑ **Auto Relief**

Rectangular

0.5

Rip Parameters

Change Direction

G   0.01in

Verify the Corners

Zoom in on the corners of the part to see the result of Insert Bends.

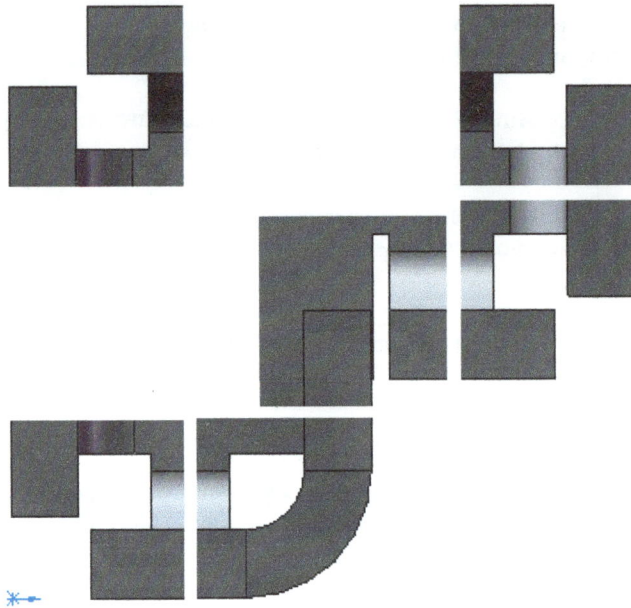

Click the **Flatten** icon in the CommandManager, or right click on **Flat-Pattern1** and pick **Unsuppress**.

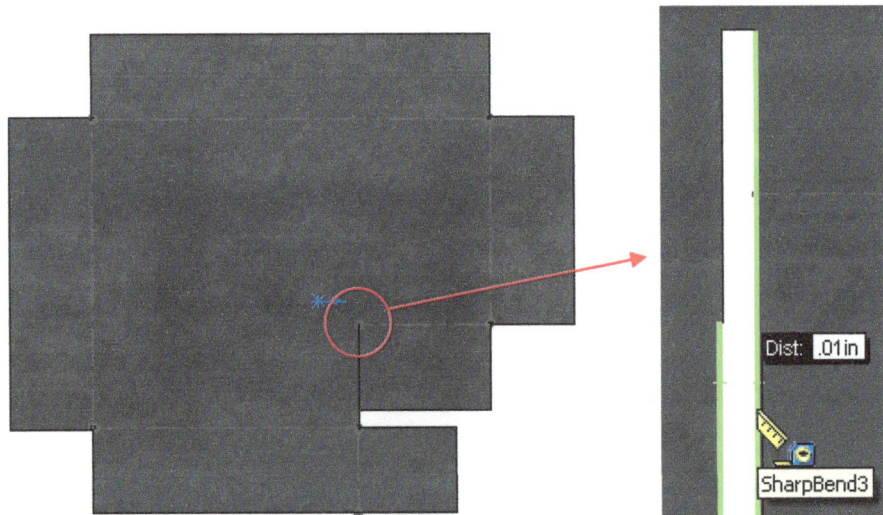

Zoom in on the notch to measure the gap size.

To do this, pull down the "Tools" menu and pick **Measure**.

Pick the two vertical lines as shown.

The **Distance** between the two lines is **0.01**.

Modify the Part

If you are using a punch press to manufacture the part, you can't punch this narrow of a slot. To fix this, a bigger gap is needed. You can edit the **Rip Gap** distance in the **Rip** PropertyManager, but the gap will be changed everywhere. Since this is undesirable, you will need to remove one edge from the original Rip feature and create a second Rip feature.

First, press the **Escape** key.

Then, click the **Flatten** icon in the CommandManager, or right click on **Flat-Pattern1** and pick **Suppress**.

In the bottom left corner of the graphics area, change the View orientation by clicking the pull down arrow and picking **Trimetric**.

In the FeatureManager design tree, drag the Rollbar below **Rip1**, or right click on **Sheet-Metal1** and pick **Rollback**.

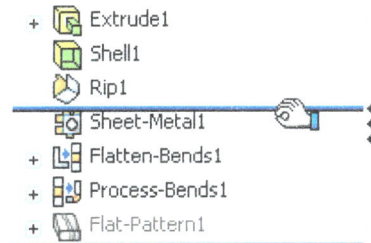

Then, in the FeatureManager design tree, right click on **Rip1** and pick **Edit Feature**.

In the graphics area, click on the notch edge line shown to deselect it (pick the line not the arrow). You can also click on the name of the edge in the list, example: **Edge<2>** (Make sure that the notch corner is the highlighted edge) under **Rip Parameters**, and press the **Delete** key to remove the edge from the **Edges to Rip** list.

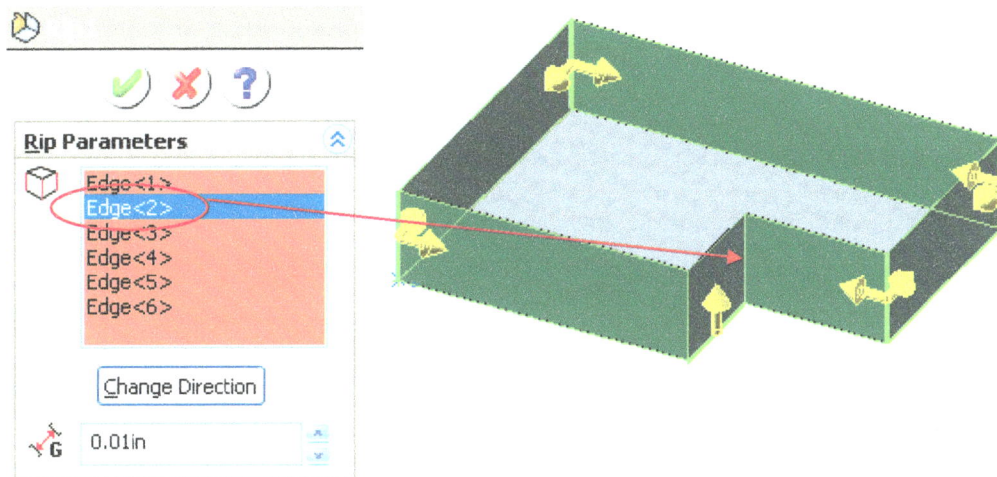

Click the green check mark button at the top of the **Rip1** PropertyManager.

Add a New Rip Feature

Click the **Rip** icon in the CommandManager, or pull down the "Insert" menu and pick **Sheet Metal - Rip**.

In the graphics area, click on the edge as shown, and then, click the right arrow.

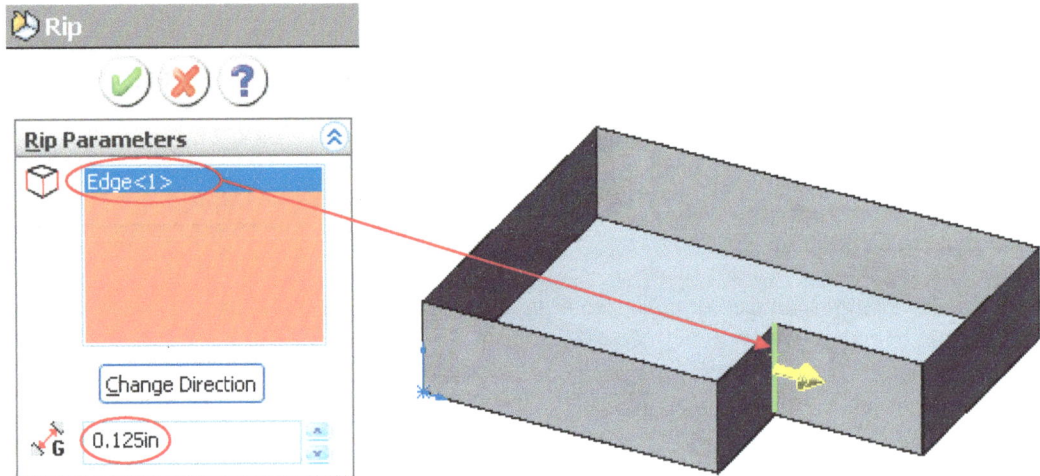

In the **Rip** PropertyManager, change the **Rip Gap** is to '**.125**'.

Click the green check mark button at the top of the **Rip** PropertyManager.

In the FeatureManager design tree, right click on **Rip2** and pick **Roll to End**.

Click the **Flatten** icon in the CommandManager, or right click on **Flat-Pattern1** and pick **Unsuppress**.

In the bottom left corner of the graphics area, change the View orientation by clicking the pull down arrow and picking **Top**, or press **Ctrl+5**.

Select the **Zoom to Area** icon to zoom in on the notch to see how it looks now.

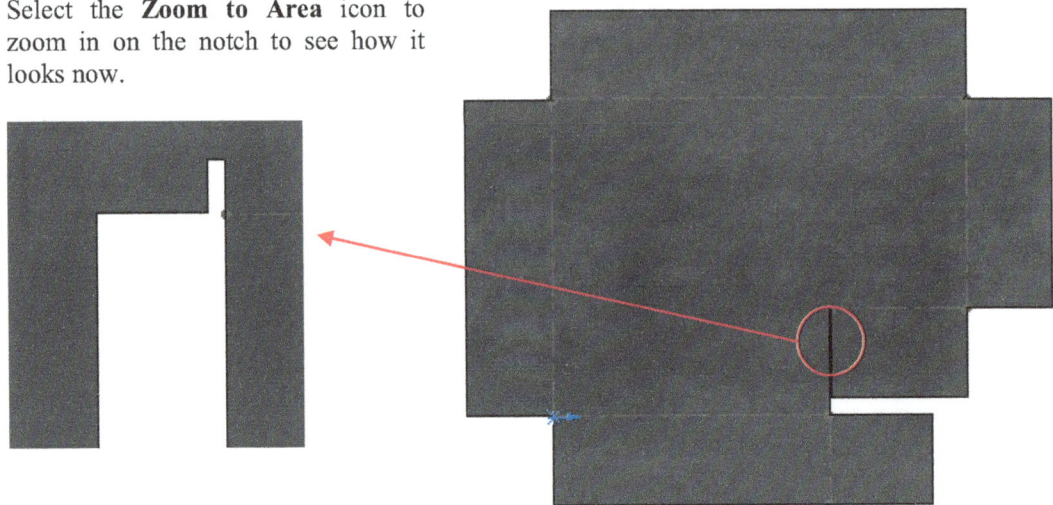

Add Manual Relief

The auto relief left a small cut out. You can add an extruded cut to manually control the relief cut to match your manufacturing tools.

Click the **Flatten** icon in the CommandManager, or right click on **Flat-Pattern1** and pick **Suppress**.

Click the **Unfold** icon in the CommandManager, or pull down the "Insert" menu and pick **Sheet Metal – Unfold**.

In the graphics area, select the top of the part for the **Fixed face**.

In the **Unfold** PropertyManager, click in the **Bends to Unfold** box.

Select the horizontal bend area as shown below.

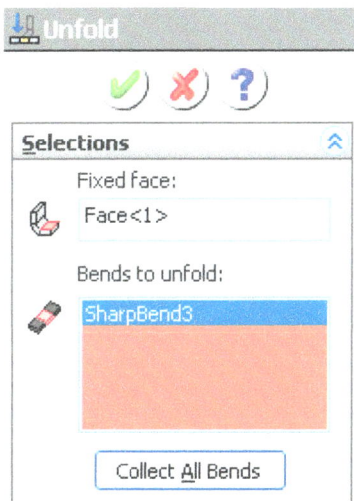

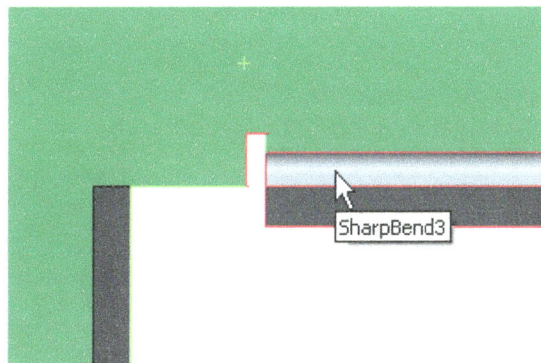

Click the green check mark button at the top of the **Unfold** PropertyManager.

Click the **Extruded Cut** icon from the CommandManager or pull down the "Insert" menu and pick **Cut – Extrude**.

Select the top of the part for the plane to sketch on.

Zoom in closer if you need to in order to pick the correct positions.

Click the **Rectangle** icon in the CommandManager, or pull down the "Tools" menu and pick **Sketch Entities – Rectangle**.

Click in the upper right corner of the relief and drag the cursor down and to the left. Place the cursor over the left edge of relief, so that it highlights, and release the mouse to create the rectangle.

Exit the sketch by clicking the **Exit Sketch** icon in the CommandManager or in the upper right corner of the graphics area.

In the **Cut-Extrude** PropertyManager, check the **Link to thickness** check box and then click the green check mark button at the top of the **Cut-Extrude** PropertyManager.

Click the **Fold** icon in the CommandManager, or pull down the "Insert" menu and pick **Sheet Metal – Fold**.

In the **Fold** PropertyManager, click on the **Collect All Bends** button to select all the appropriate bends in the part.

Click the green check mark button at the top of the **Fold** PropertyManager to fold the part back up.

Click the **Flatten** icon in the CommandManager, or right click on **Flat-Pattern1** and pick **Unsuppress**.

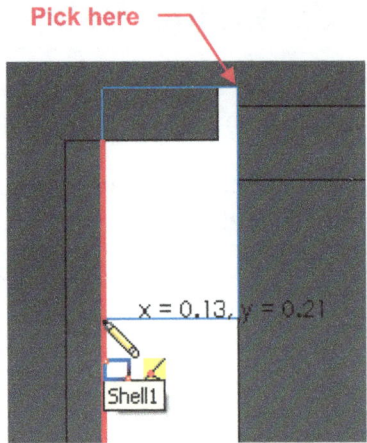

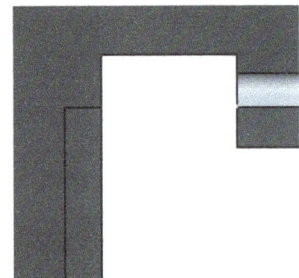

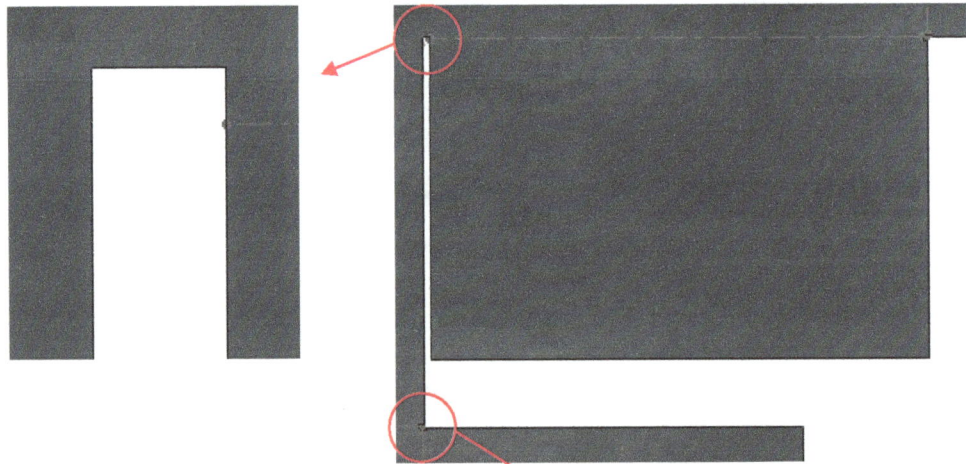

Check Corner Reliefs

Now, you will check the other corner reliefs and adjust them for manufacturing.

Zoom in to one of the corner reliefs.

Pull down the "Tools" menu and pick **Measure**.

Pick the left vertical line as shown. The **Length** of the line is **0.104in**.

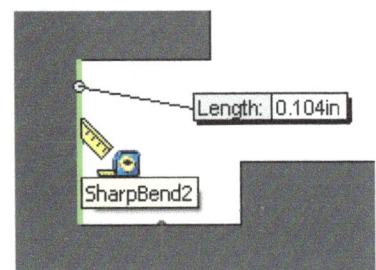

Add a Corner Trim Feature

Again, to ensure that a standard tool size can be used to notch out the corners, a **Corner Trim** will be used. For this example, assume that you will be punching the part and are planning to clean the notches out with a 1/8 square punch. Note that the **Corner Trim** tool cuts or adds material only to flattened sheet metal parts.

Press the **Escape** key, and then, pull down the "Insert" menu and pick **Sheet Metal – Corner Trim**.

In the **Corner-Trim** PropertyManager, click the **Collect all Corners** button. This selects all six corners. You will remove the inside corner in a moment. You could just as easily manually selected the five corners in the graphics area.

Set the **Relief Type** to **Square**.

Check **Centered on bend lines** to center the cut.

Enter a **Side length** of '**.125**'.

Click the **Zoom to Fit** icon in the "View" toolbar, or pull down the "View" menu and pick **Modify – Zoom to Fit**.

In the graphics area, pick the corner shown to deselect it, removing it from the **Corner edges** list. You may need to zoom in to be able to deselect the corner. There should be a total of five corners selected.

Click the green check mark button at the top of the **Corner Trim** PropertyManager.

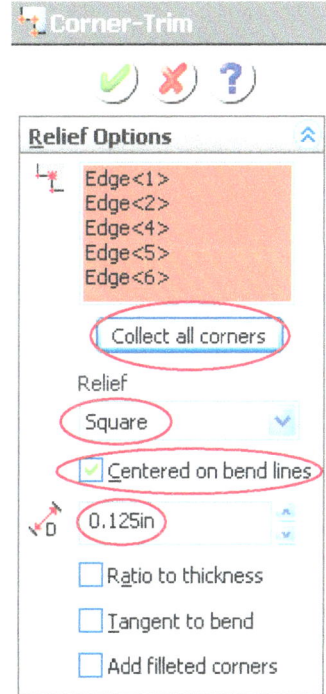

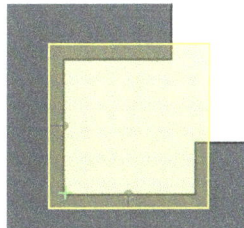

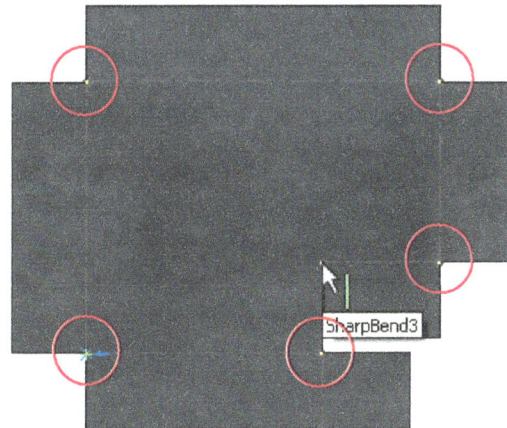

Zoom in to one of the corner reliefs.

Pull down the "Tools" menu and pick **Measure**.

Pick the left vertical line as shown.

The **Length** of the line is now **0.125in**.

Press the **Escape** key.

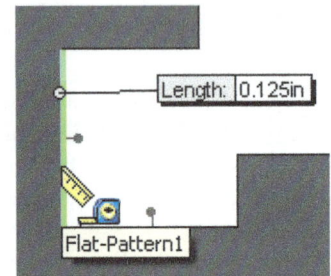

Click the **Flatten** icon in the CommandManager, or right click on **Flat-Pattern1** and pick **Suppress**.

In the bottom left corner of the graphics area, change the View orientation by clicking the pull down arrow and picking **Trimetric**.

Saving the Part

Click the **Save** icon in the "Standard" toolbar, or pick **Save** from the "File" pull down menu.

In the **Save As** dialog box, in the **File name** box, type '**Chapter 8**' and click **Save**.

Chapter 9

Mounting Clip

Large radius bends can present some unique situations, especially when there is not a flange in between the bend areas.

Creating the part as a cross section is easy, but creating a part this way locks in a single value for use on all bend areas in the cross section. Even with simple 90 degree bends and the use of Bend Allowance values, this would give the wrong flat, since the bends have different radii.

Modeling the part to separate the bends to independent features is more difficult, but it provides the control required to specify the correct bend value for each bend area.

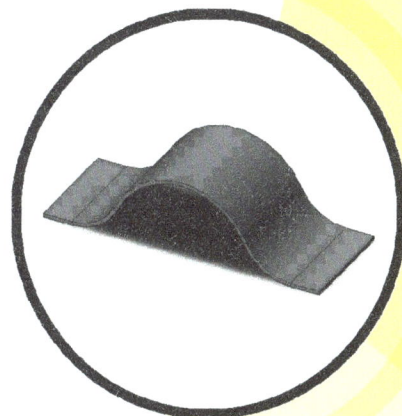

Create the Base Flange

Begin a new **Part** document by clicking the **New** icon in the "Standard" toolbar, or pull down the "File" menu and pick **New**.

In the FeatureManager design tree, right click on **Material**, and pick **1060 Alloy** from the menu.

Create a base flange by clicking the **Sheet Metal** icon in the control area of the CommandManager. Then, click the **Base-Flange/Tab** icon from the toolbar, or pull down the "Insert" menu and pick **Sheet Metal – Base Flange**.

Select the **Front** plane when prompted to select a plane on which to sketch the feature cross-section.

Create a small horizontal line to the left of the origin using the **Line** icon in the CommandManager, or pull down the "Tools" menu and pick **Sketch Entities – Line**.

Then, click on the **Tangent Arc** icon in the CommandManager, or pull down the "Tools" menu and pick **Sketch Entities – Tangent Arc**.

Create a tangent arc starting at the right endpoint of the line and curving up. Then, create another arc up and to the right. Finally, create a third arc down to the right, ending horizontal to the original line as shown below.

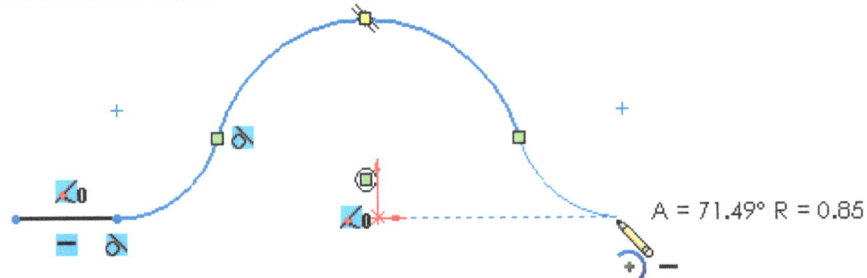

Then, create another small horizontal line starting at the end of the last tangent arc to the right of the origin using the **Line** icon in the CommandManager, or pull down the "Tools" menu and pick **Sketch Entities – Line**.

Press the **Escape** key. **Ctrl** select the center point of the center arc and the origin. Then, in the **Properties** PropertyManager, click the **Coincident** button.

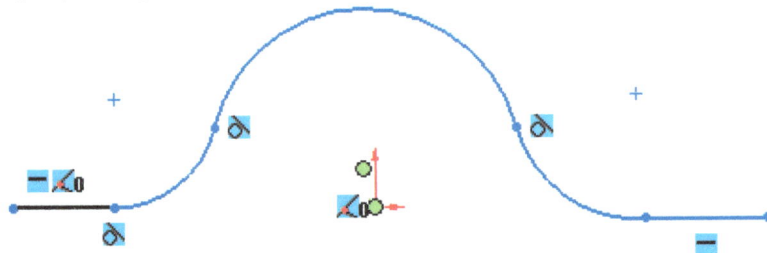

Press the **Escape** key. If they don't already have the tangent relation associated to them, **Ctrl** select the small arc on the right and the right horizontal line. Then, in the **Properties** PropertyManager, click the **Tangent** button.

Press the **Escape** key. **Ctrl** select the small arc on the left and the small arc on the right. Then, in the **Properties** PropertyManager, click the **Equal** button.

Press the **Escape** key. **Ctrl** select the left horizontal line and the right horizontal line. Then, in the **Properties** PropertyManager, click the **Equal** button and the **Collinear** button.

Add the dimensions as shown using the **Smart Dimension** icon in the CommandManager, or pull down the "Tools" menu and pick **Dimensions – Smart**.

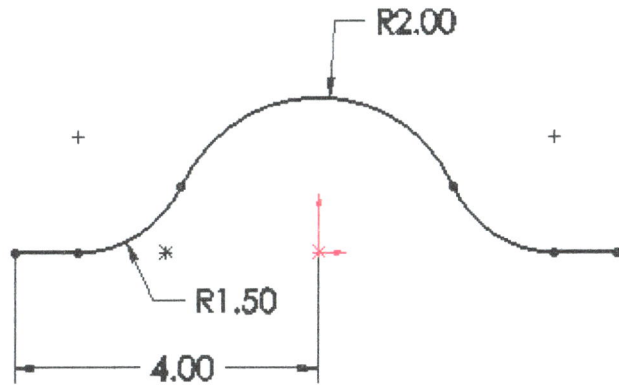

Exit the sketch by clicking the **Exit Sketch** icon in the CommandManager or in the upper right corner of the graphics area.

In the **Base Flange** PropertyManager, under **Direction 1**, set the **End Condition** to **Blind** and the **Depth** to '3'.

Under **Sheet Metal Gauges**, check the **Use gauge table** check box.

Pick **ALUM GAUGES** from the **Select Table** pull down.

Under **Sheet Metal Parameters**, set the thickness to **10 Gauge**.

Make sure that the **Reverse direction** check box is checked.

Due to the large radius bends in this part, you will need a different bend factor than the default value shown. Under **Bend Allowance**, check the **Override value** check box. Enter a **K-Factor** of '.42'. This value will be used for all three bends on the part.

If you want the **2.00** radius bend to use a different bend parameter than the **1.50** radius bends, you'll have to use the k-factor table method. This style of creating the part is the easiest, but it restricts the application of bend values to one number fits all.

Click the green check mark button at the top of the **Base Flange** PropertyManager to accept the settings and create the part.

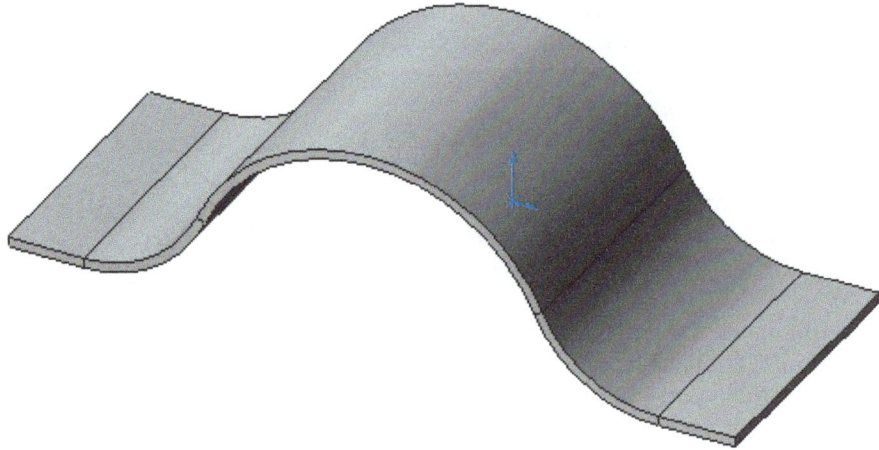

To unfold it, simply click the **Flatten** icon in the CommandManager, or right click on **Flat-Pattern1** and pick **Unsuppress**.

Select the top of the part. Then, in the bottom left corner of the graphics area, change the View orientation by clicking the pull down arrow and picking **Normal To**.

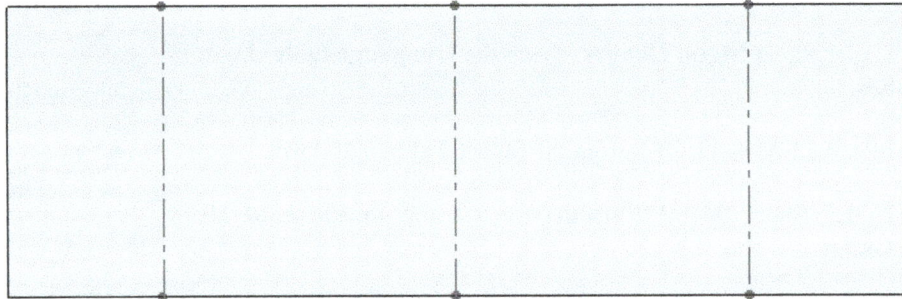

Click the **Flatten** icon in the CommandManager, or right click on **Flat-Pattern1** and pick **Suppress**.

In the bottom left corner of the graphics area, change the View orientation by clicking the pull down arrow and picking **Trimetric**.

Saving the Part

Click the **Save** icon in the "Standard" toolbar, or pick **Save** from the "File" pull down menu.

In the **Save As** dialog box, in the **File name** box, type '**Clip**' and click **Save**.

Another Way of Drawing the Part

Begin a new **Part** document by clicking the **New** icon in the "Standard" toolbar, or pull down the "File" menu and pick **New**.

In the FeatureManager design tree, right click on **Material**, and pick **1060 Alloy** from the menu.

Create a base flange by clicking the **Sheet Metal** icon in the control area of the CommandManager. Then, click the **Base-Flange/Tab** icon from the toolbar, or pull down the "Insert" menu and pick **Sheet Metal – Base Flange**.

Select the **Front** plane when prompted to select a plane on which to sketch the feature cross-section.

Create an arc using the **Centerpoint Arc** icon in the CommandManager, or pull down the "Tools" menu and pick **Sketch Entities – Centerpoint Arc**.

Pick the origin as the center point for the arc and then move the cursor up and to the right to pick a starting point. Move the cursor to the left of the origin and pick the end point.

Press the **Escape** key. **Ctrl** select the end points of the arc. Then, in the **Properties** PropertyManager, click the **Horizontal** button.

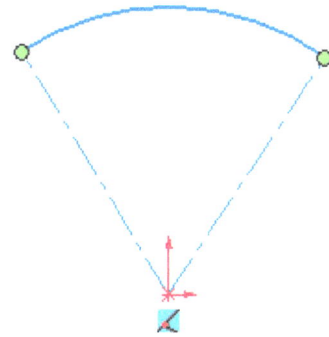

Create a centerline line from each end of the arc to the origin using the **Centerline** icon in the CommandManager, or pull down the "Tools" menu and pick **Sketch Entities – Centerline**.

Add a '2.0' radius dimension as shown using the **Smart Dimension** icon in the CommandManager, or pull down the "Tools" menu and pick **Dimensions – Smart**.

Then, add a '129.25' angular dimension between the two centerlines. Yes, this is an odd number, but that is what the first part ended up as.

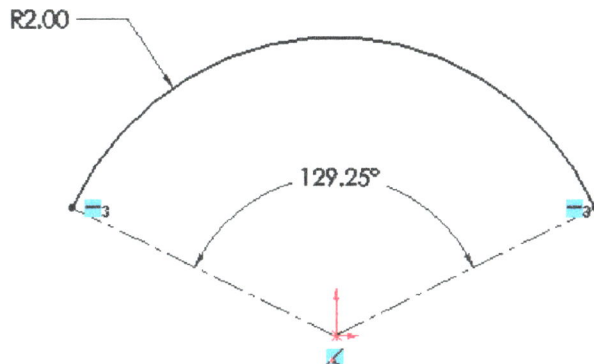

Exit the sketch by clicking the **Exit Sketch** icon in the CommandManager or in the upper right corner of the graphics area.

In the **Base Flange** PropertyManager, under **Direction 1**, set the **End Condition** to **Blind** and the **Depth** to '**3**'.

Under **Sheet Metal Gauges**, check the **Use gauge table** check box.

Pick **ALUM GAUGES** from the **Select Table** pull down.

Under **Sheet Metal Parameters**, set the **Thickness** to **10 Gauge**.

Make sure that the **Reverse direction** check box is checked.

Under **Bend Allowance**, check the **Override value** check box.

Enter '**0.45**' for the **K-Factor**.

Click the green check mark button at the top of the **Base Flange** PropertyManager to accept the settings and create the part.

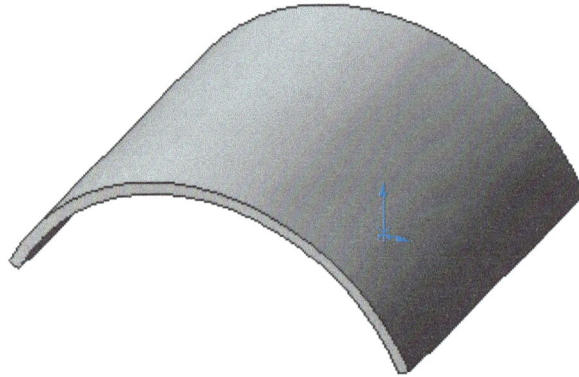

Add an Edge Flange

Click the **Edge Flange** icon in the CommandManager, or pull down the "Insert" menu and pick **Sheet Metal – Edge Flange**.

Pick the upper right edge of the part and move the cursor up and to the right to show the direction of the flange. Make sure that you create a long flange for the next step.

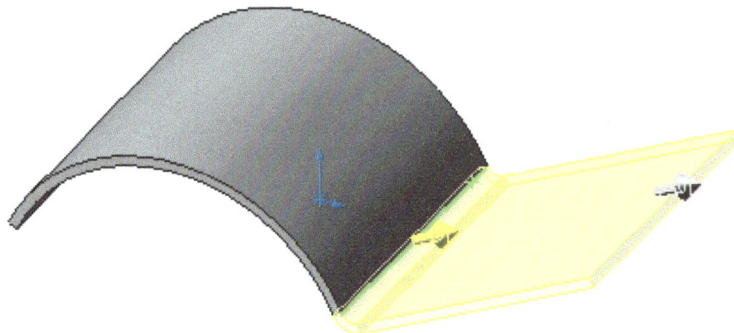

In the **Edge Flange** PropertyManager, under **Flange Parameters**, uncheck the **Use default radius** check box, and set the **Bend Radius** to '**1.5**'.

Under **Angle**, set the **Flange Angle** to '**64.625**'.

Under **Flange Length**, set the **Length End Condition** to **Blind** and the **Length** to '**2.0**'. Then, click the **Outer Virtual Sharp** icon and the **Bend Outside Flange Position** icon.

Check the **Custom Bend Allowance** check box, and set the **K-Factor** to '**.42**'.

Click the green check mark button at the top of the **Edge-Flange** PropertyManager to accept the settings and create the flange.

Repeat the previous steps on the other side of the Base Flange. Don't forget to set the K-Factor value. If you try to mirror this flange rather than create it a second time, SolidWorks will fail to create it, most likely due to not having a flat in between the base and edge flanges.

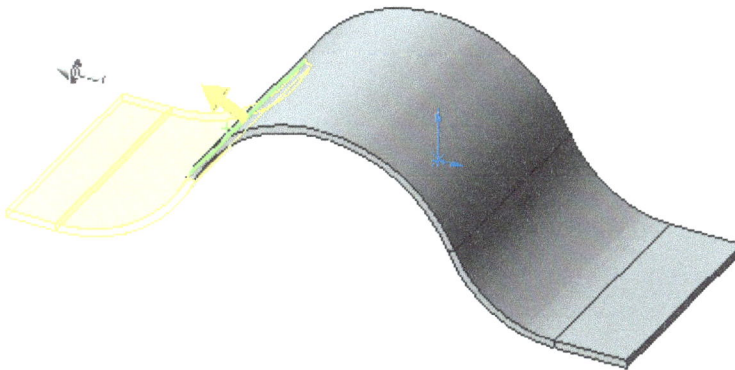

Click the **Flatten** icon in the CommandManager, or right click on **Flat-Pattern1** and pick **Unsuppress**.

Select the top of the part. Then, in the bottom left corner of the graphics area, change the View orientation by clicking the pull down arrow and picking **Normal To**.

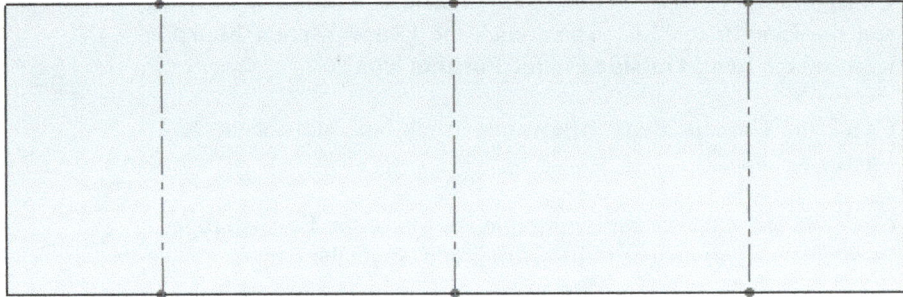

Click the **Flatten** icon in the CommandManager, or right click on **Flat-Pattern1** and pick **Suppress**.

In the bottom left corner of the graphics area, change the View orientation by clicking the pull down arrow and picking **Trimetric**.

Saving the Part

Click the **Save** icon in the "Standard" toolbar, or pick **Save** from the "File" pull down menu.

In the **Save As** dialog box, in the **File name** box, type 'Clip1' and click **Save**.

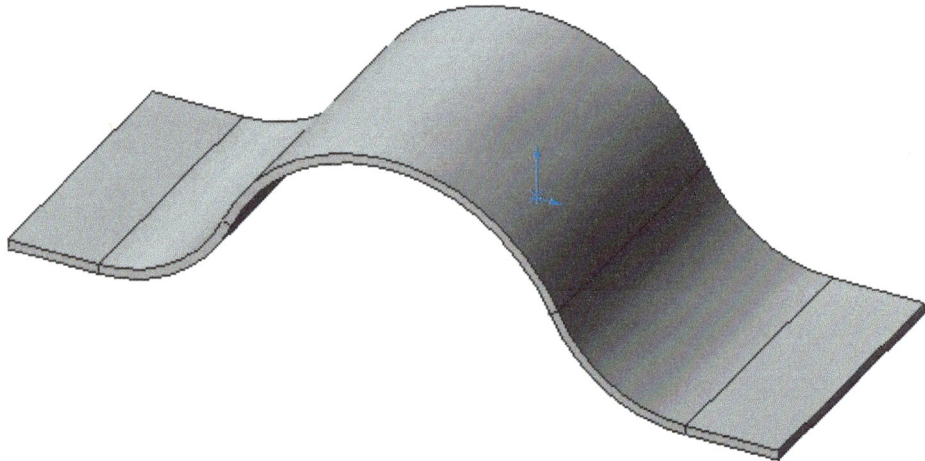

Chapter 10

Airplane

Many times parts must be imported from a DXF or other file type. The geometry may or may not be good. This is something you will have to validate for each part.

To properly fold the part to the correct 3D model you will also need all of the bend information which was used when the part was originally created. This includes the bend radius and the exact material thickness.

Sometimes the order of bends will cause SolidWorks to not be able to fold the part. Just try fewer bends in each step and change the order until it works.

This part comes from a few years back. One of the machine tool builders was demonstrating their equipment at a trade show by cutting the parts and bending them on their brake. We took one and reverse engineered it. So, let's have some fun.

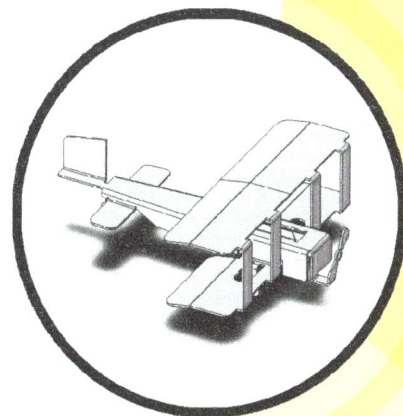

Download the DXF File

Go to http://www.sheetmetalguy.com/ and select **Downloads** from the navigation menu. This will take you to the correct page to download the file **airplane.dxf**.

Save **airplane.dxf** in your SolidWorks directory.

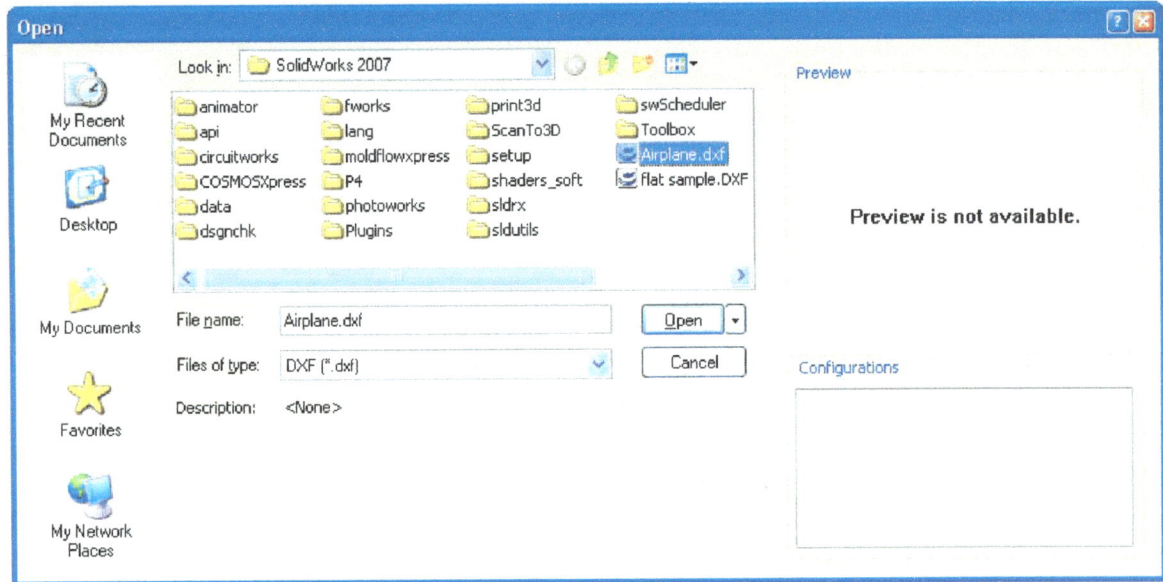

Import the DXF File

Click the **Open** icon in the "Standard" toolbar, or pull down the "File" menu and pick **Open**.

In the **Open** dialog box, change the **Files of type** to **DXF (*.dxf)**.

Explore to the SolidWorks directory where you saved the downloaded airplane.dxf file.

Click on **airplane.dxf** and click the **Open** button.

In the **DXF/DWG Import** dialog box, select **Import to a new part**, and click **Next**.

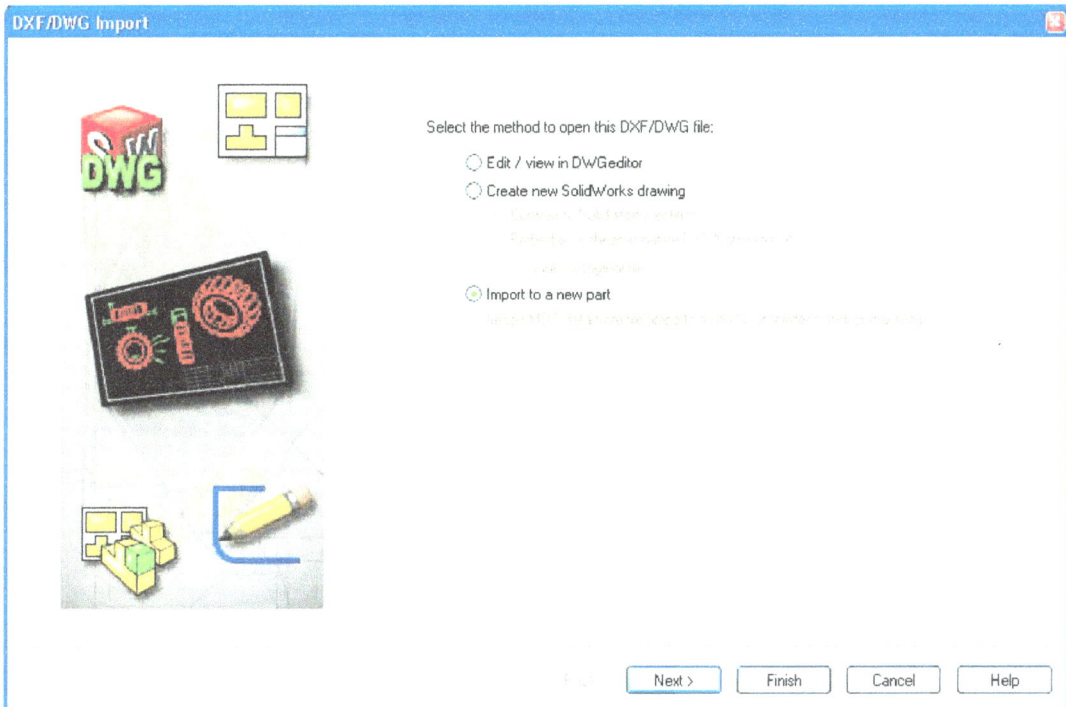

In the **DXF/DWG Import – Drawing Layer Mapping** dialog box, click **Next**.

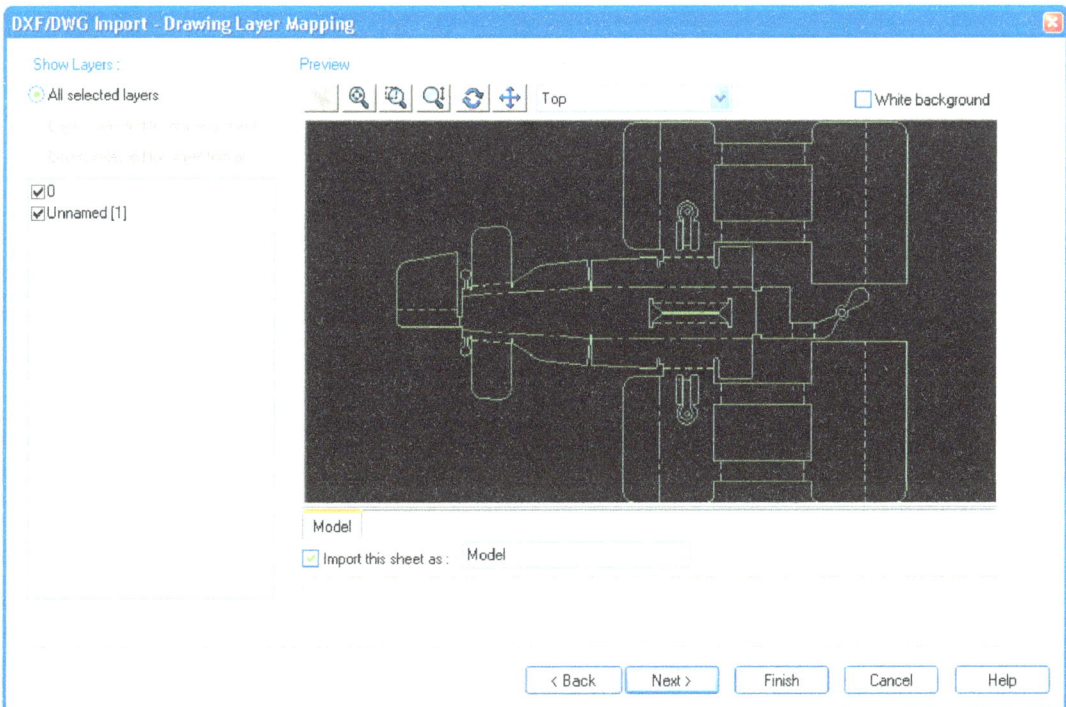

In the **DXF/DWG Import – Document Settings** dialog box, click **Finish**.

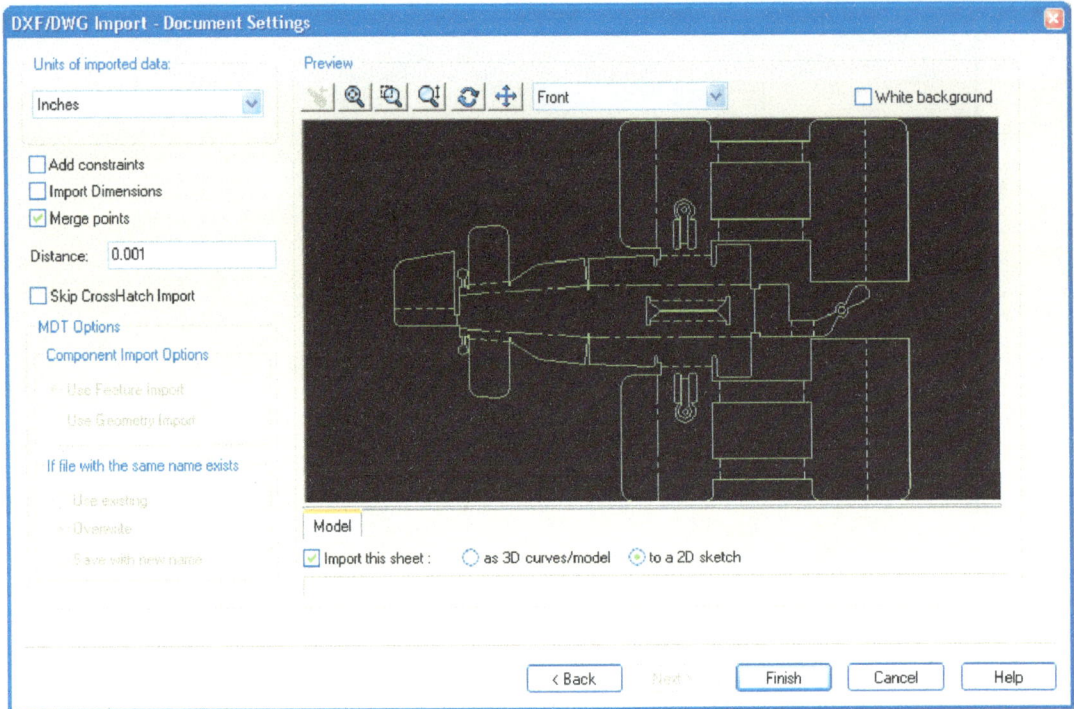

Create a base flange by clicking the **Sheet Metal** icon in the control area of the CommandManager. Then, click the **Base-Flange/Tab** icon from the toolbar, or pull down the "Insert" menu and pick **Sheet Metal – Base Flange**.

In the **Base Flange** PropertyManager, under **Sheet Metal Gauges**, check the **Use gauge table** check box.

Pick **CRS GAUGES** from the **Select Table** pull down.

Under **Sheet Metal Parameters**, set the **Thickness** to **18 Gauge**.

Make sure that the **Reverse direction** check box is not checked.

Click the green check mark button at the top of the **Base Flange** PropertyManager to accept the settings and create the part.

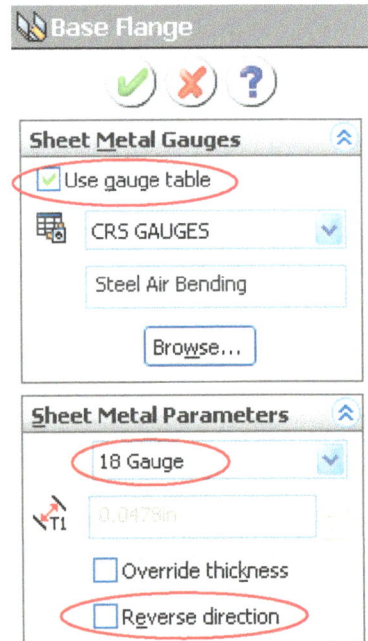

In the FeatureManager design tree, right click on **Sheet-Metal1**, and pick **Edit Feature** from the menu.

In the **Sheet-Metal1** PropertyManager, under **Sheet Metal Parameters**, make sure that the **Bend Radius** is set to **0.05in**.

Under **Bend Allowance**, check the **Override value** check box and enter a **K-Factor** of '**.42**'.

Click the green check mark button at the top of the **Sheet-Metal1** PropertyManager.

In the FeatureManager design tree, right click on **Material**, and pick **Plain Carbon Steel** from the menu.

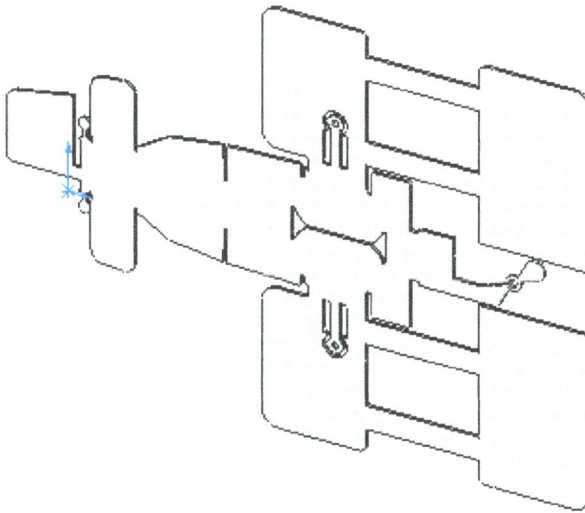

In the FeatureManager design tree, click on the plus sign next to **Base-Flange1**.

Right click on **Sketch1** and pick **Edit Sketch Plane** from the menu.

In the upper left hand corner of the graphics area, click on the plus sign next to **Part1** to expand the flyout FeatureManager design tree.

In the flyout FeatureManager design tree, click on **Top Plane**.

Click the green check mark button at the top of the **Sketch Plane** PropertyManager.

Bend the Top Wing Flap

In the FeatureManager design tree, right click on **Sketch1** and pick **Show** from the menu.

Click the **Sheet Metal** icon in the control area of the CommandManager. Then, click the **Sketched Bend** icon from the toolbar, or pull down the "Insert" menu and pick **Sheet Metal – Sketched Bend**.

Select the top of the part when prompted to select a planar face on which to sketch the bend lines.

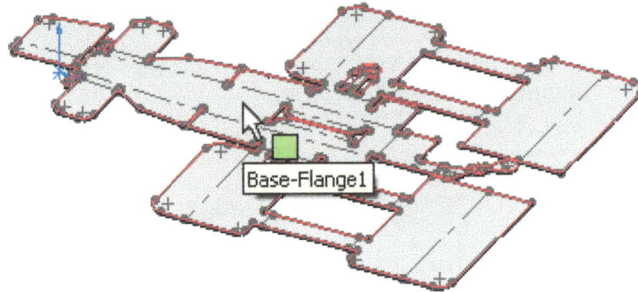

Ctrl select the two lines shown, and then click the **Convert Entities** icon, or pull down the "Tools" and pick **Sketch Tools – Convert Entities**.

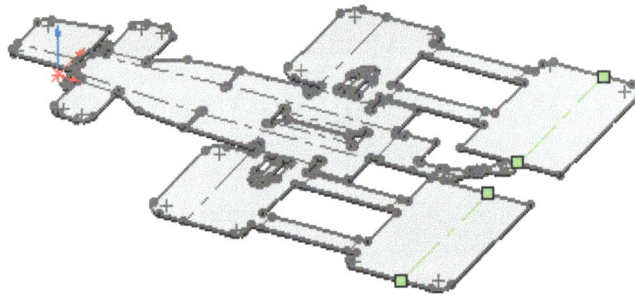

Exit the sketch by clicking the **Exit Sketch** icon in the CommandManager or in the upper right corner of the graphics area.

In the graphics area, select the top of the part as shown.

In the **Sketched Bend** PropertyManager, make sure that the **Bend Centerline** button is selected.

Enter '**10**' for the **Bend Angle**.

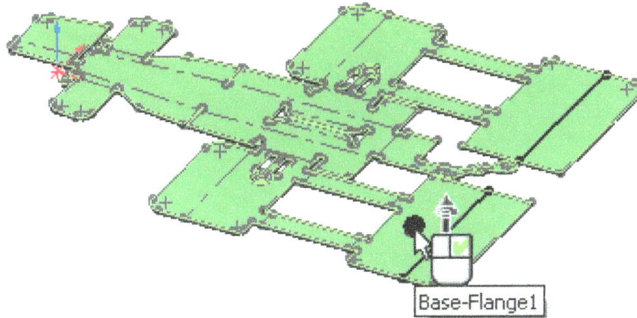

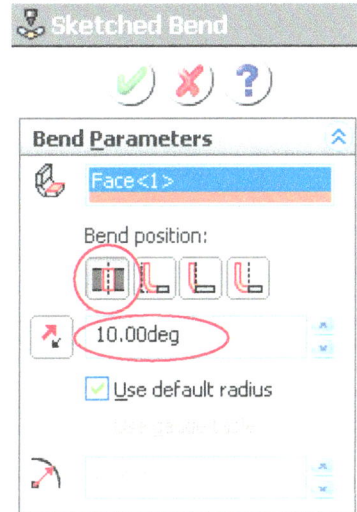

Click the green check mark button at the top of the **Sketched Bend** PropertyManager.

Bend the Bottom Wing Flap

Click the **Sketched Bend** icon from the CommandManager, or pull down the "Insert" menu and pick **Sheet Metal – Sketched Bend**.

Select the top of the part.

Ctrl select the two lines shown, and then click the **Convert Entities** icon, or pull down the "Tools" and pick **Sketch Tools – Convert Entities**.

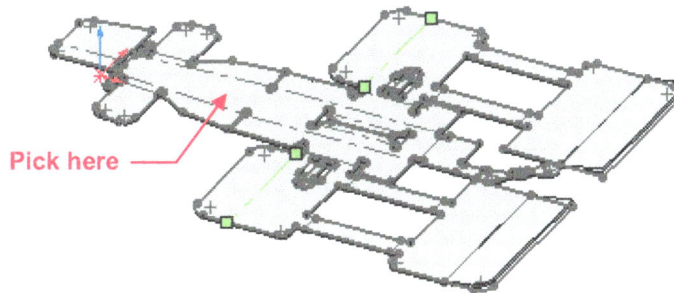

Pick here

Exit the sketch by clicking the **Exit Sketch** icon in the CommandManager or in the upper right corner of the graphics area.

In the graphics area, select the top of the part as shown.

Enter '**10**' for the **Bend Angle**.

Click the **Reverse Direction** button. The arrow on the part should point down.

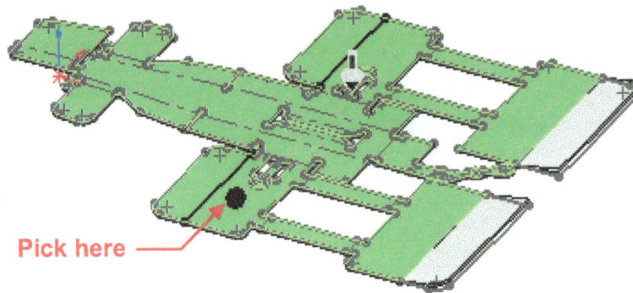

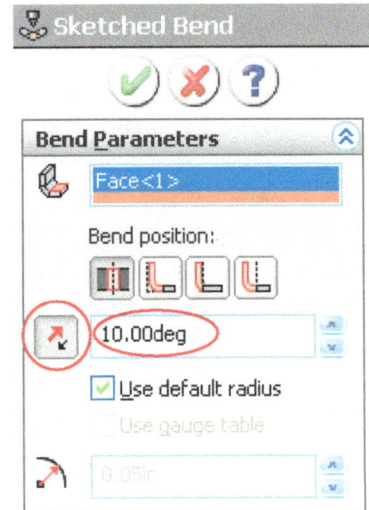

Pick here

✓   Click the green check mark button at the top of the **Sketched Bend** PropertyManager.

Bend the Wheels

Click the **Sketched Bend** icon from the CommandManager, or pull down the "Insert" menu and pick **Sheet Metal – Sketched Bend**.

Select the top of the part.

Ctrl select the two lines shown, and then click the **Convert Entities** icon, or pull down the "Tools" and pick **Sketch Tools – Convert Entities**. You may need to zoom in to select the correct lines.

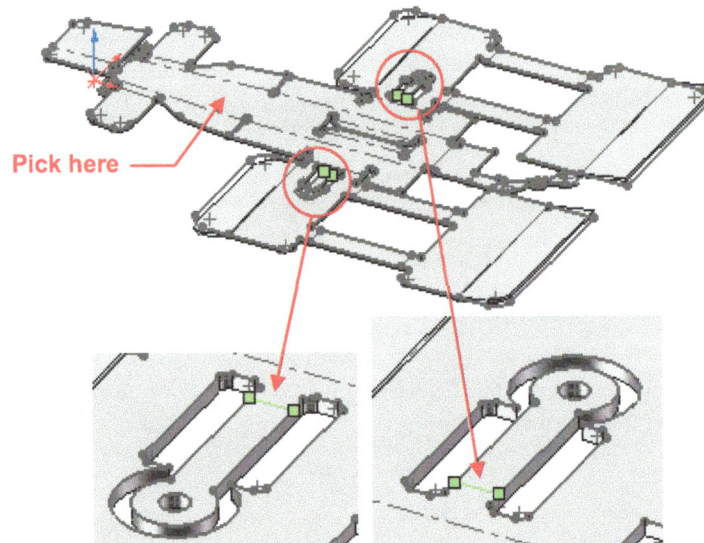

Pick here

Exit the sketch by clicking the **Exit Sketch** icon in the CommandManager or in the upper right corner of the graphics area.

In the graphics area, select the top of the part as shown.

Enter '**90**' for the **Bend Angle**.

Click the **Reverse Direction** button. The arrow on the part should point down.

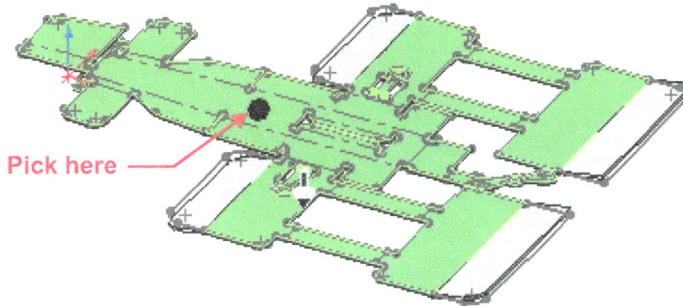

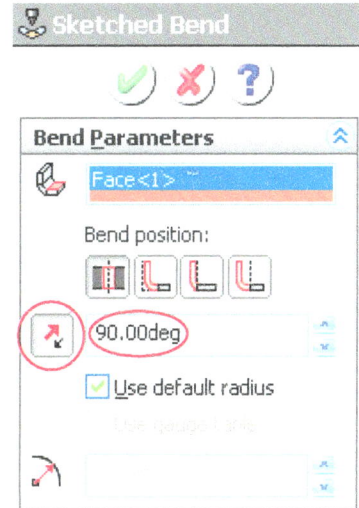

Pick here

✓ Click the green check mark button at the top of the **Sketched Bend** PropertyManager.

Bend the Cockpit

Click the **Sketched Bend** icon from the CommandManager, or pull down the "Insert" menu and pick **Sheet Metal – Sketched Bend**.

Select the top of the part.

Ctrl select the two lines shown, and then click the **Convert Entities** icon, or pull down the "Tools" and pick **Sketch Tools – Convert Entities**.

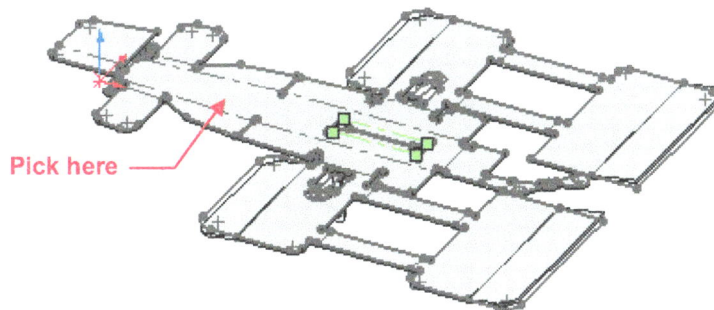

Pick here

Exit the sketch by clicking the **Exit Sketch** icon in the CommandManager or in the upper right corner of the graphics area.

In the graphics area, select the top of the part as shown.

Enter '45' for the **Bend Angle**.

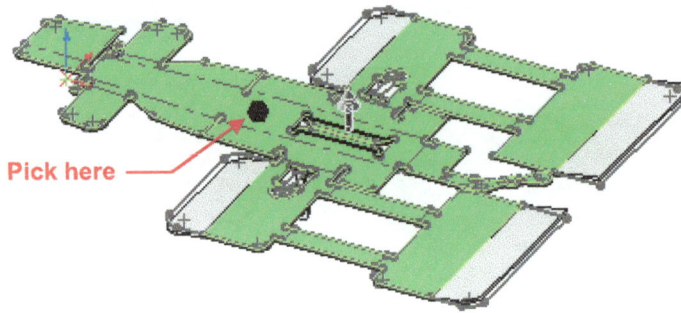

Pick here

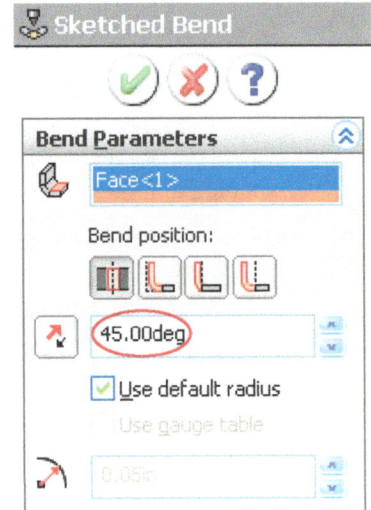

Sketched Bend

Bend Parameters

Face<1>

Bend position:

45.00deg

☑ Use default radius

Use gauge table

0.05in

Click the green check mark button at the top of the **Sketched Bend** PropertyManager.

Bend the Propeller

Click the **Sketched Bend** icon from the CommandManager, or pull down the "Insert" menu and pick **Sheet Metal – Sketched Bend**.

Select the top of the part.

Select the two lines shown, and then click the **Convert Entities** icon, or pull down the "Tools" and pick **Sketch Tools – Convert Entities**. You may need to zoom in to select the correct lines.

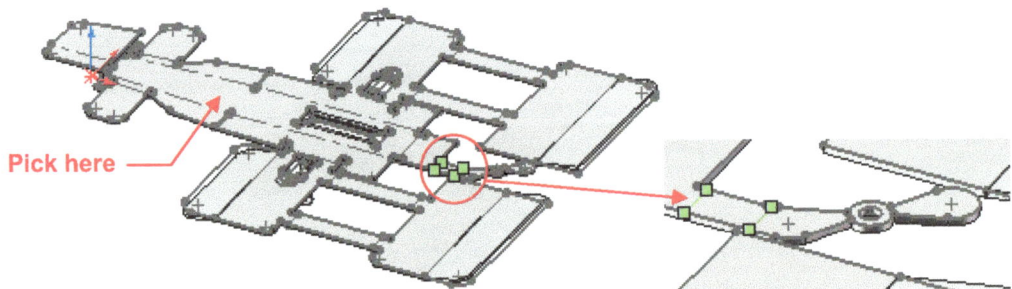

Pick here

Exit the sketch by clicking the **Exit Sketch** icon in the CommandManager or in the upper right corner of the graphics area.

In the graphics area, select the top of the part as shown.

Enter '**90**' for the **Bend Angle**.

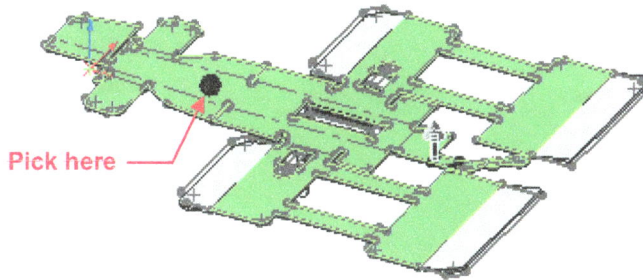

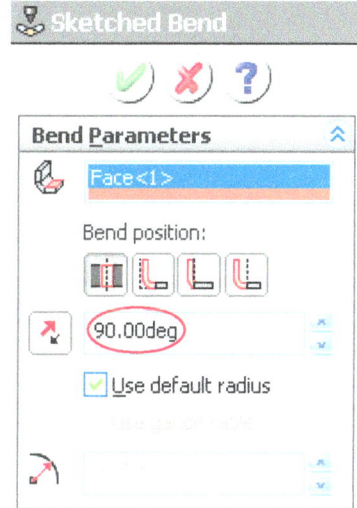

Pick here

Sketched Bend

Bend Parameters

Face<1>

Bend position:

90.00deg

☑ Use default radius

✓ Click the green check mark button at the top of the **Sketched Bend** PropertyManager.

Click the **Sketched Bend** icon from the CommandManager again, or pull down the "Insert" menu and pick **Sheet Metal – Sketched Bend**.

Select the top of the part.

Select the line shown, and then click the **Convert Entities** icon, or pull down the "Tools" and pick **Sketch Tools – Convert Entities**.

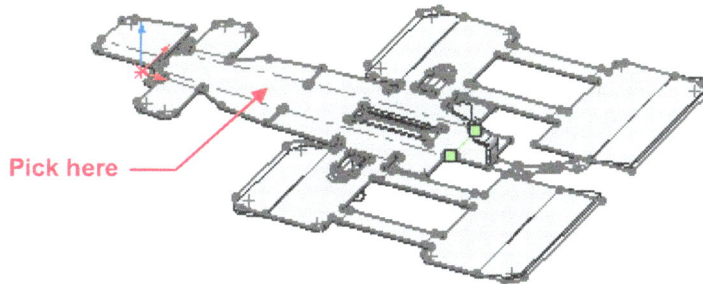

Pick here

Exit the sketch by clicking the **Exit Sketch** icon in the CommandManager or in the upper right corner of the graphics area.

In the graphics area, select the top of the part as shown.

Enter '90' for the **Bend Angle**.

Click the **Reverse Direction** button. The arrow on the part should point down.

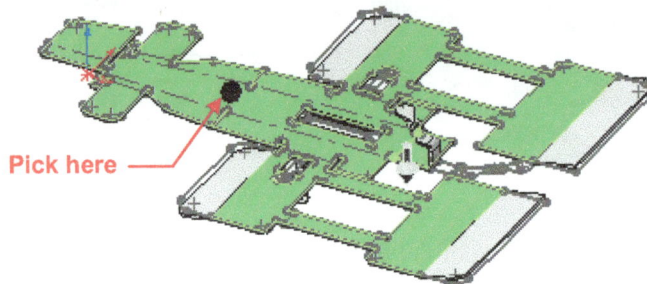

Pick here

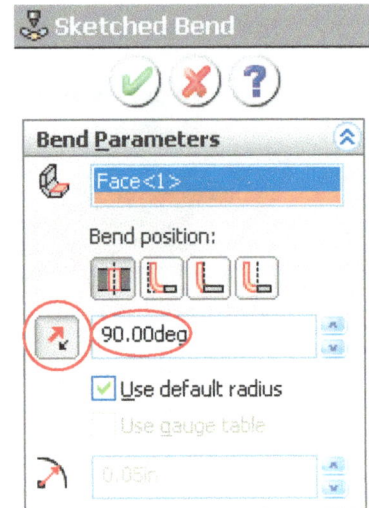

Click the green check mark button at the top of the **Sketched Bend** PropertyManager.

Bend the Wings

Click the **Sketched Bend** icon from the CommandManager, or pull down the "Insert" menu and pick **Sheet Metal – Sketched Bend**.

Select the top of the part.

Select the five lines shown, and then click the **Convert Entities** icon, or pull down the "Tools" and pick **Sketch Tools – Convert Entities**. You may need to zoom in to select the correct lines.

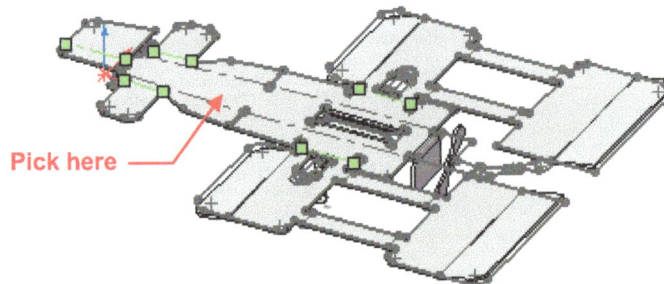

Pick here

Exit the sketch by clicking the **Exit Sketch** icon in the CommandManager or in the upper right corner of the graphics area.

In the graphics area, select the top of the part as shown.

Enter '**90**' for the **Bend Angle**.

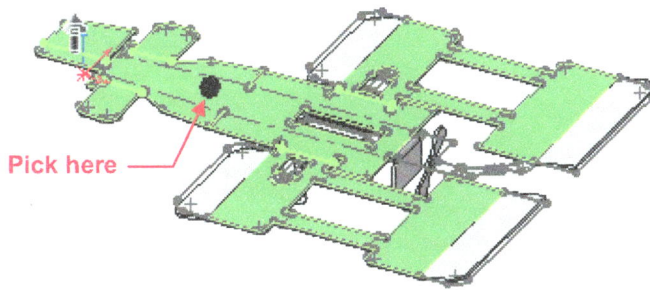

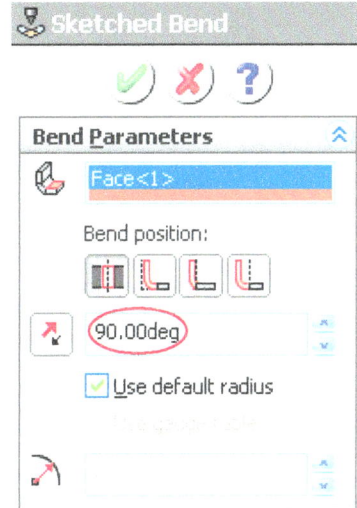

Pick here

Sketched Bend

Bend Parameters

Face<1>

Bend position:

90.00deg

☑ Use default radius

✓ Click the green check mark button at the top of the **Sketched Bend** PropertyManager.

⚖ Click the **Sketched Bend** icon from the CommandManager, or pull down the "Insert" menu and pick **Sheet Metal – Sketched Bend**.

Select the top of the part.

▢ Select the four lines shown, and then click the **Convert Entities** icon, or pull down the "Tools" and pick **Sketch Tools – Convert Entities**. You may need to zoom in to select the correct lines.

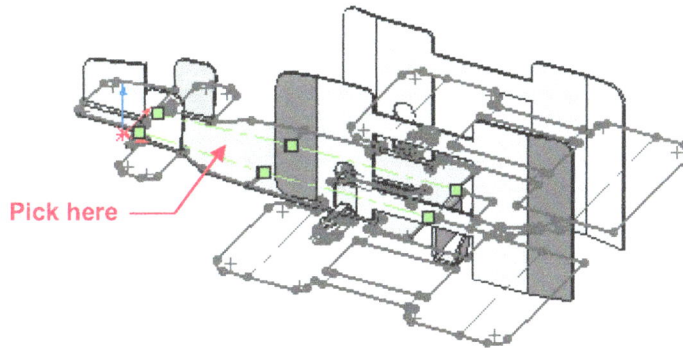

Pick here

✏ Exit the sketch by clicking the **Exit Sketch** icon in the CommandManager or in the upper right corner of the graphics area.

In the graphics area, select the top of the part as shown.

Enter '**90**' for the **Bend Angle**.

Click the **Reverse Direction** button. The arrow on the part should point down.

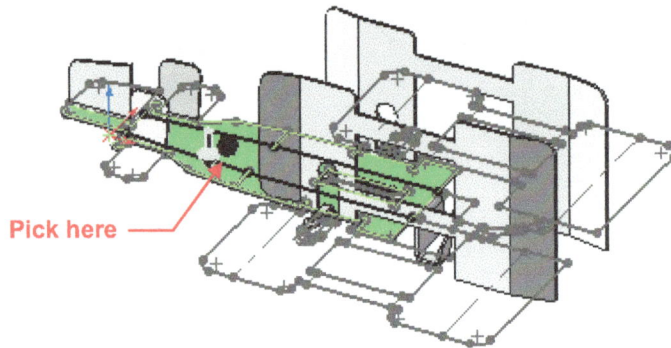

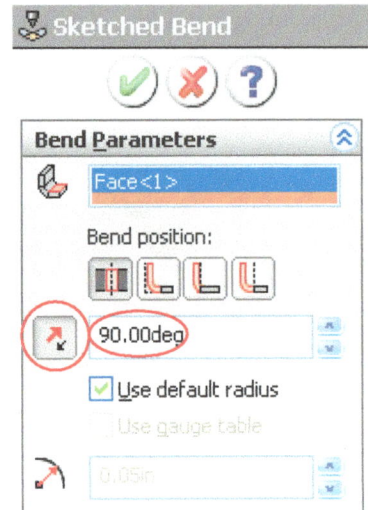

Pick here

Click the green check mark button at the top of the **Sketched Bend** PropertyManager.

Bend the Main Wings

Click the **Sketched Bend** icon from the CommandManager, or pull down the "Insert" menu and pick **Sheet Metal – Sketched Bend**.

Select the top of the part on a portion of the wing which has not already been bent.

In the bottom left corner of the graphics area, change the View orientation by clicking the pull down arrow and picking **Top**, or press **Ctrl+5**.

Select the four lines in the sketch as shown, and then click the **Convert Entities** icon, or pull down the "Tools" and pick **Sketch Tools – Convert Entities**. You may need to zoom in to select the correct lines.

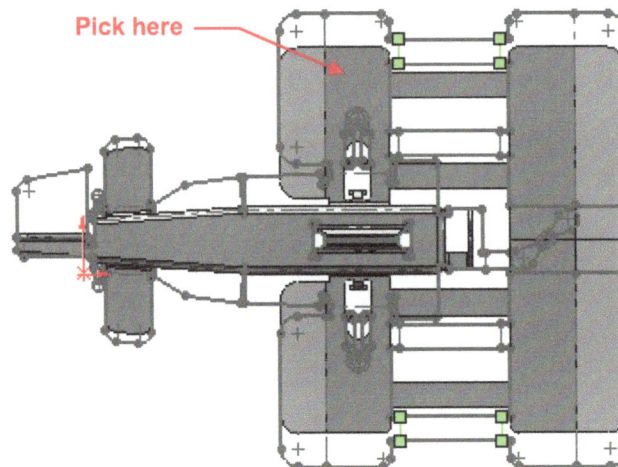

Pick here

Left click and drag the endpoints of the small vertical lines to go across the wing sections as shown below.

Exit the sketch by clicking the **Exit Sketch** icon in the CommandManager or in the upper right corner of the graphics area.

In the graphics area, select the top of the wing as shown.

Enter '**90**' for the **Bend Angle**.

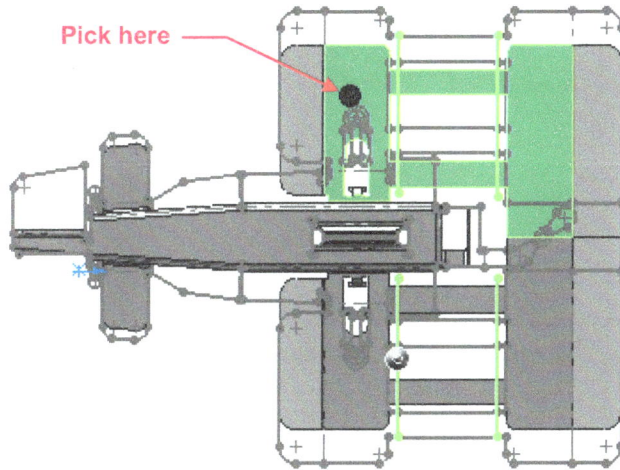

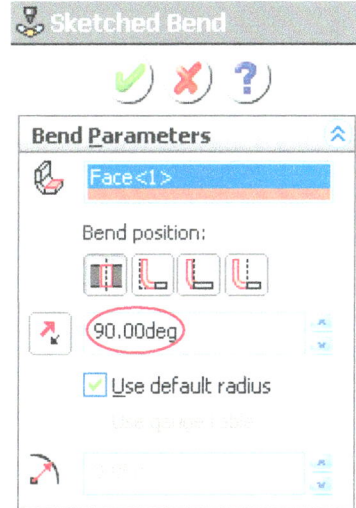

Pick here

Click the green check mark button at the top of the **Sketched Bend** PropertyManager.

In the bottom left corner of the graphics area, change the View orientation by clicking the pull down arrow and picking **Trimetric**.

In the FeatureManager design tree, right click on **Sketch1** and pick **Hide** from the menu.

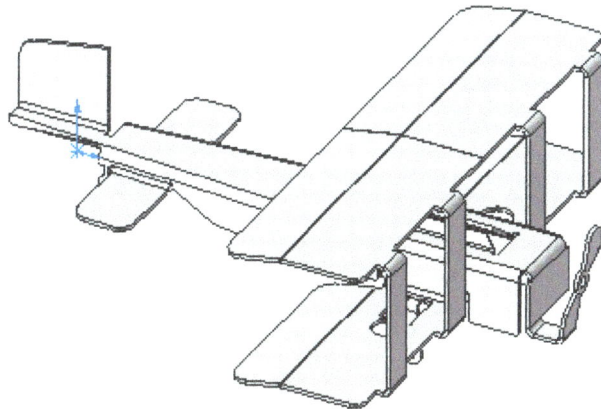

Save the Part

Click the **Save** icon in the "Standard" toolbar, or pick **Save** from the "File" pull down menu.

The **Save As** dialog box appears. In the **File name** box, type '**Airplane**' and click **Save**.

Fix the Incorrect Bend Radius

You may have noticed that the side bend of the plane did not bend properly. A partial bend appears on the tail of the plane that should not have a bend on it.

To fix this, in the FeatureManager design tree, right click on **Sheet-Metal1** and pick **Edit Feature** from the menu.

In the **Sheet-Metal1** PropertyManager, under **Sheet Metal Parameters**, check the **Override radius** check box and enter a **Bend Radius** of '.03'.

Click the green check mark button at the top of the **Sheet-Metal1** PropertyManager.

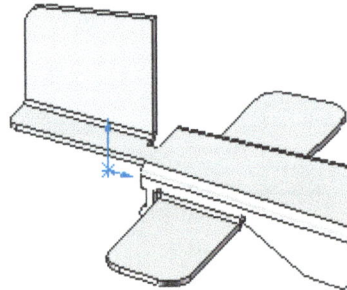

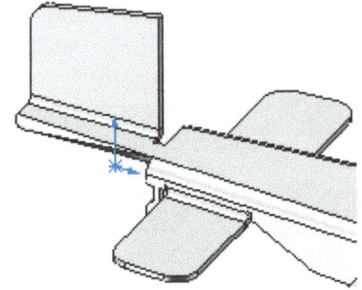

Fix the Wing Overlap

Now that the tail of the plane is correct, the wings slightly overlap. To figure out why this is so, you can measure the part.

To do this, in the bottom left corner of the graphics area, change the View orientation by clicking the pull down arrow and picking **Right**, or press **Ctrl+4**.

Pull down the "Tools" menu and pick **Measure**.

Select the left and right vertical edges of the center of the plane.

At first glance the plane appears to have the desired width of 1in.

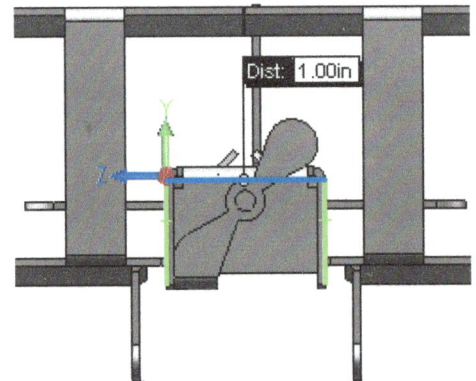

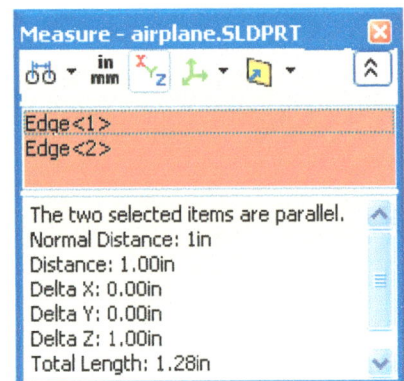

in
mm

In the **Measure** dialog box, click the **Units/Precision** icon.

In the **Measure Units/Precision** dialog box, set the **Decimal places** to '**4**' and click **OK**.

The true width is actually **0.9997in**.

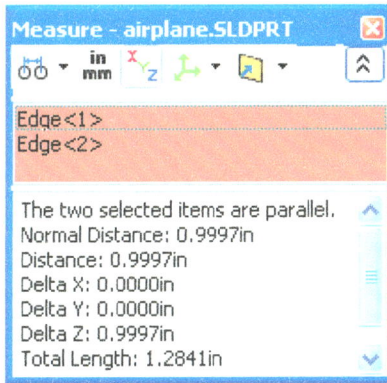

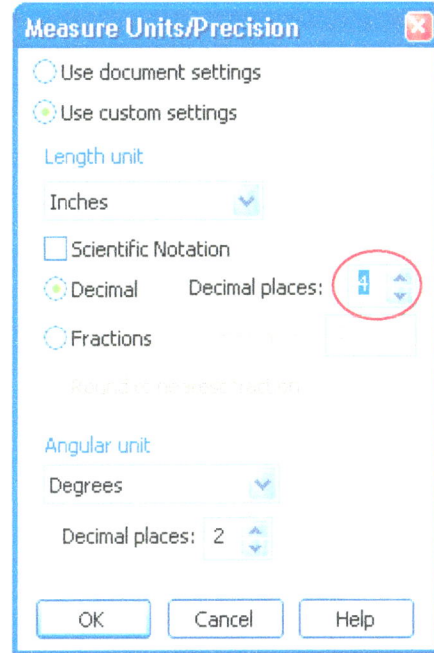

The selected gauge table uses a sheet metal thickness of 0.0478 for 18 gauge material. Try changing the thickness of the airplane to 0.048.

To do this, press the **Escape** key. Then, in the FeatureManager design tree, right click on **Sheet-Metal1** and pick **Edit Feature** from the menu.

In the **Sheet-Metal1** PropertyManager, under **Sheet Metal Parameters**, check the **Override thickness** check box and enter a **Thickness** of '**.048**'.

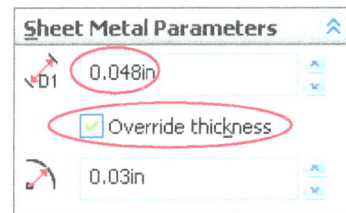

Click the green check mark button at the top of the **Sheet-Metal1** PropertyManager.

Pull down the "Tools" menu and pick **Measure**.

Select the left and right vertical edges of the center of the plane.

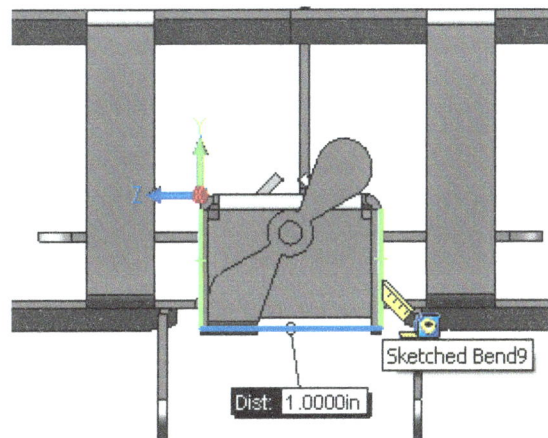

The true width is now **1.0000in**., and the wings no longer overlap.

Press the **Escape** key.

In the bottom left corner of the graphics area, change the View orientation by clicking the pull down arrow and picking **Trimetric**.

Make sure that the bends are valid by clicking the **Sheet Metal** icon in the control area of the CommandManager. Then, click the **Flatten** icon from the toolbar. You may also right click on **Flat-Pattern1** and pick **Unsuppress**.

If the part flattens without any errors, the part is good.

Click the **Flatten** icon from the CommandManager. You may also right click on **Flat-Pattern1** and pick **Suppress**.

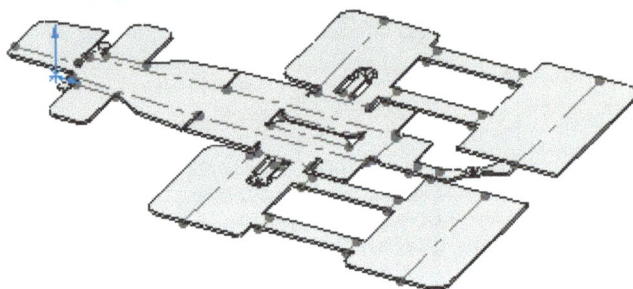

Save the Part

Click the **Save** icon in the "Standard" toolbar, or pick **Save** from the "File" pull down menu.

The **Save As** dialog box appears. In the **File name** box, type '**Airplane**' and click **Save**.

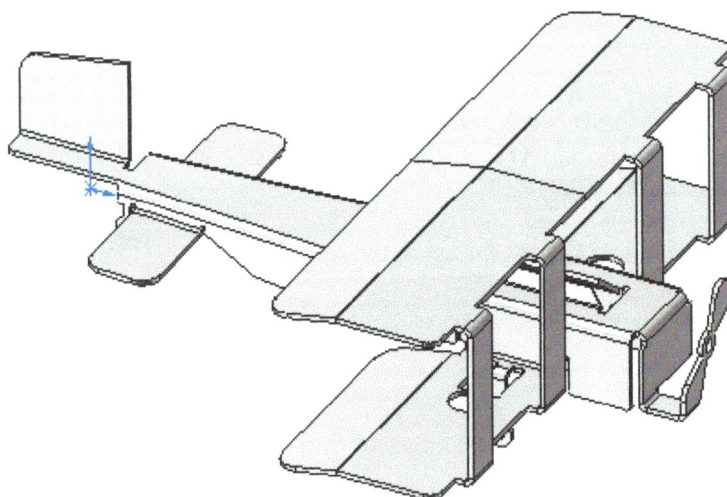

Chapter 11

Converting Parts to Sheet Metal

SolidWorks can open files created in other formats for viewing and editing purposes. The standard Open command is used to open files in any of the available formats. This chapter uses an IGES file format created from the cover part file in Chapter 6 of Course 2.

However, SolidWorks does not recognize the part as sheet metal and you must use the Insert bend command in order to unfold the part. Your ability to modify the part is then also limited as you will see.

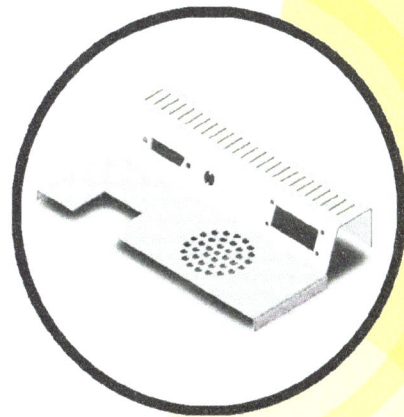

Download the IGES File

Go to http://www.sheetmetalguy.com/ and select **Downloads** from the navigation menu to take you to the download page.

Download the file **cover.igs** and save the file in your SolidWorks directory.

Import the IGES File

Click the **Open** icon in the "Standard" toolbar, or pull down the "File" menu and pick **Open**.

In the **Open** dialog box, change the **Files of type** to **IGES (*.igs;*.iges)**.

Explore to the SolidWorks directory where you saved the downloaded cover.igs file.

Click on **cover.igs** and click the **Open** button.

In the **Import Diagnostics** dialog box, click **Yes**. If the **Import Diagnostics** dialog box does not appear, right click on **Imported1** in the FeatureManager design tree and pick **Import Diagnostics**.

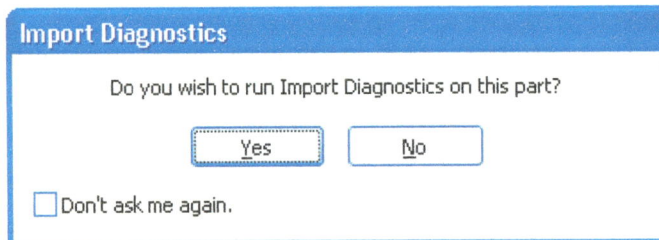

In the **Import Diagnostics** PropertyManager, click on the **Attempt to Heal All** button.

If SolidWorks still shows an error with faulty faces, just continue. The model has been tested on various versions and service packs to make certain that the instructions in this chapter will work properly.

Click the green check mark button at the top of the **Import Diagnostics** PropertyManager.

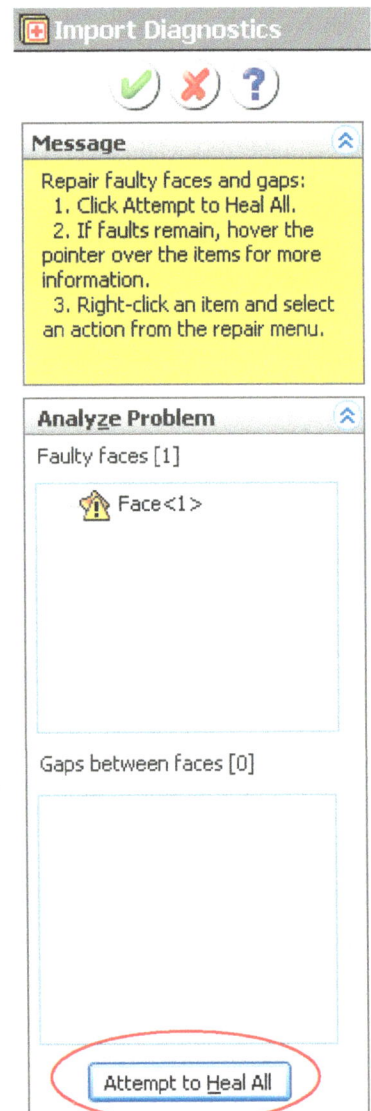

Convert the Part to Sheet Metal

In the FeatureManager design tree, right click on **Material**, and pick **Plain Carbon Steel** from the menu.

Use the **Insert Bends** command to get SolidWorks to identify the part as sheet metal and associate the bend parameters to the bend areas.

Click the **Sheet Metal** icon in the control area of the CommandManager. Then, click the **Insert Bends** icon from the toolbar, or pull down the "Insert" menu and pick **Sheet Metal - Bends**.

Select the top of the lower side of the part.

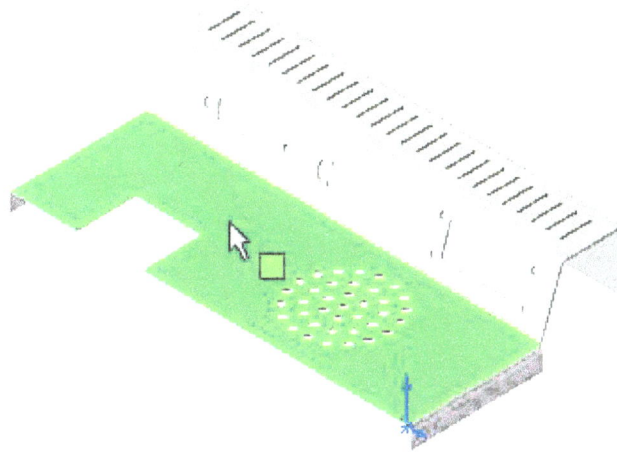

In the **Bends** PropertyManager, under **Bend Parameters**, enter '.075' for the **Bend Radius**. This is a default radius. SolidWorks will find the radius modeled in each bend area and use that as the bend radius for that bend area.

Under **Bend Allowance**, set the **K-Factor** to '.333'.

Click the green check mark button at the top of the **Bends** PropertyManager to accept the settings.

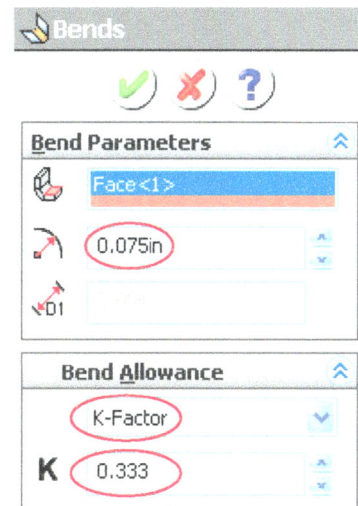

Save the Part

Click the **Save** icon in the "Standard" toolbar, or pick **Save** from the "File" pull down menu.

The **Save As** dialog box appears. In the **File name** box, type '**Cover**' and click **Save**.

Review the Results

In the FeatureManager design tree, click on the plus sign to the left of **Flatten-Bends1**. The five bend areas of the part are shown in the list.

Click on **RoundBend1**. In the graphics area, the related bend area is highlighted.

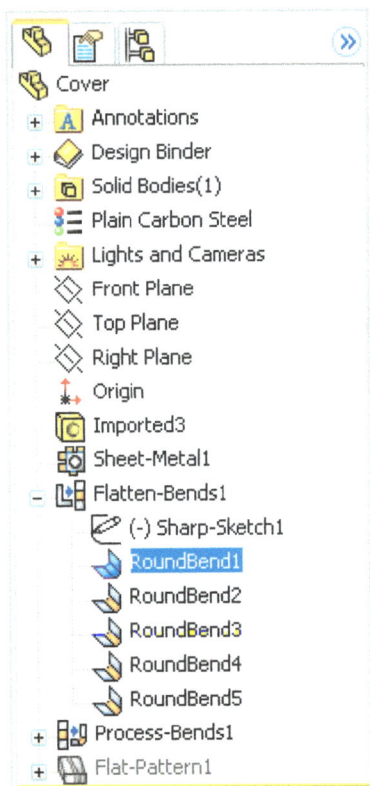

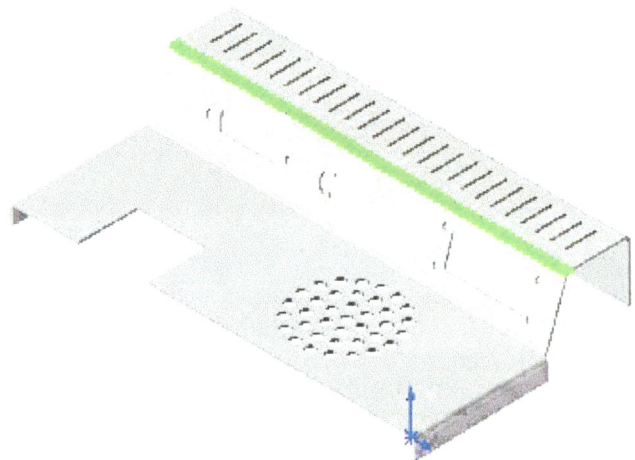

Right click on **RoundBend1** and pick **Edit Feature**. Note that the **Bend Angle** and **Bend Radius** are both grayed out. You cannot edit these items. You can, however, override the default bend allowance and specify a custom bend allowance for this bend only.

Click the green check mark button at the top of the **RoundBend1** PropertyManager.

While you were allowed to import the part and tell SolidWorks it is sheet metal, you cannot easily modify the part or the bend areas.

Unfold the Part

Click the **Sheet Metal** icon in the control area of the CommandManager. Then, click the **Flatten** icon from the toolbar. You may also right click on **Flat-Pattern1** in the FeatureManager design tree and pick **Unsuppress**. This displays the flat pattern of the part.

In the bottom left corner of the graphics area, change the View orientation by clicking the pull down arrow and picking **Top**, or press **Ctrl+5**.

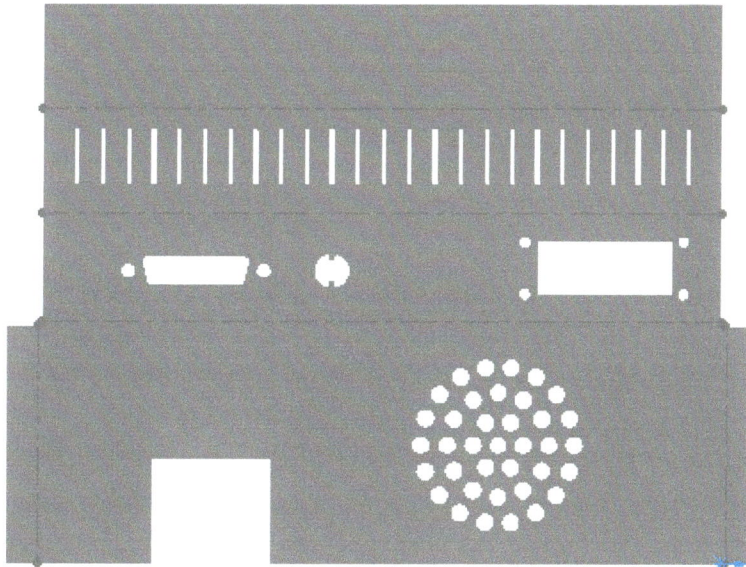

If the correct bend factor is given you will get the correct flat pattern. This is also an opportunity to be creative and lie about the bend value in order to get the correct flat pattern. While this requires some calculation on your part it can result in the correct flat pattern geometry to then be passed on to your CAM software.

Click the **Flatten** icon in the CommandManager, or right click on **Flat-Pattern1** in the FeatureManager design tree and pick **Suppress**.

In the bottom left corner of the graphics area, change the View orientation by clicking the pull down arrow and picking **Isometric**, or press **Ctrl+7**.

Save the Part

Click the **Save** icon in the "Standard" toolbar, or pick **Save** from the "File" pull down menu.

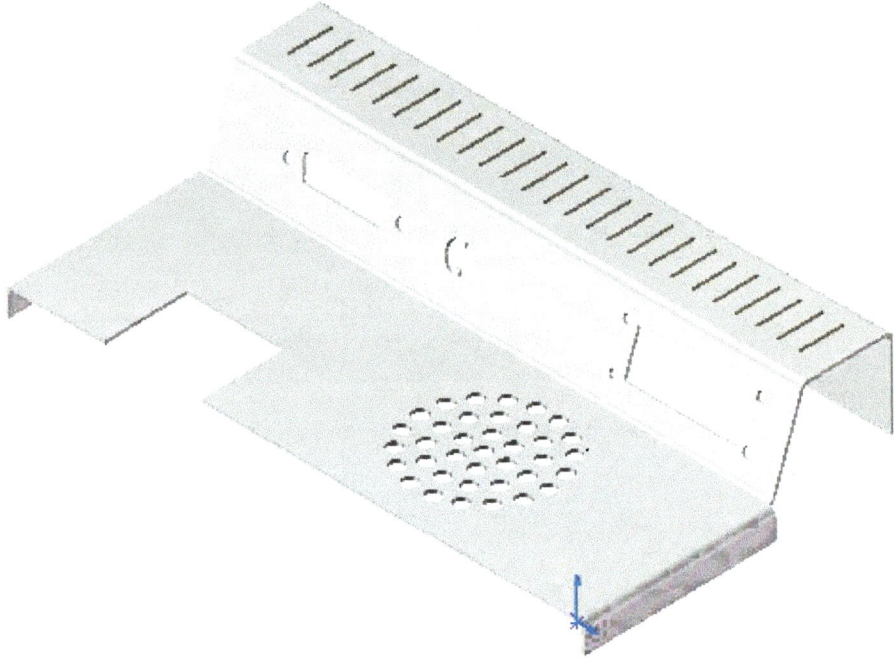

Chapter 12

Import and Convert

Now that you had some fun, let's try the real thing. You receive a file from the customer, and the part needs to be adjusted to your bend factors.

In this chapter, you will import a Parasolid file and convert it to a sheet metal part. Then, you will add the correct bend radius and parameters. Remember, this is a simple part to show you the technique. Your parts will most likely be more complex and may require more effort to get all of the parameters correct.

While this will let you at the bend areas, the holes and cutouts are not readily available for editing. That again requires more work, so hopefully your customer modeled those correctly.

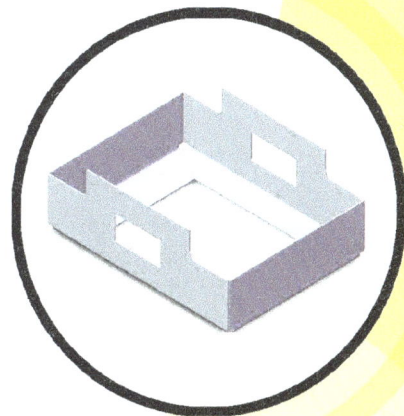

Download the Parasolid File

Go to http://www.sheetmetalguy.com/ and select **Downloads** from the Navigation menu. This will take you to the correct page to download the file **BoxImport.x_t**.

Save **BoxImport.x_t** in your SolidWorks directory.

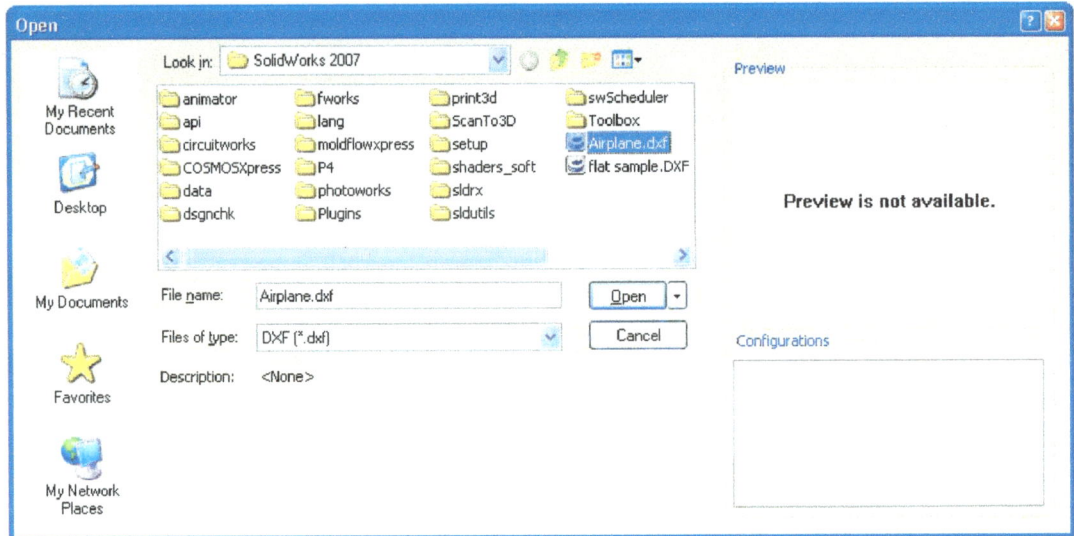

Import the Parasolid File

Click the **Open** icon in the "Standard" toolbar, or pull down the "File" menu and pick **Open**.

In the **Open** dialog box, change the **Files of type** to **Parasolid (*.x_t, *.x_b, *xmt_txt, *xmt_bin)**.

Explore to the SolidWorks directory where you saved the downloaded **BoxImport.x_t** file.

Click on **BoxImport.x_t** and click the **Open** button.

The **Import Diagnostics** dialog box appears asking, **"Do you wish to run Import Diagnostics on this part?"**

Click the **No** button.

Click the **Zoom to Area** icon in the "View" toolbar to zoom in to the bottom corner of the part and take a look at what you have. There are no arcs in the bend areas.

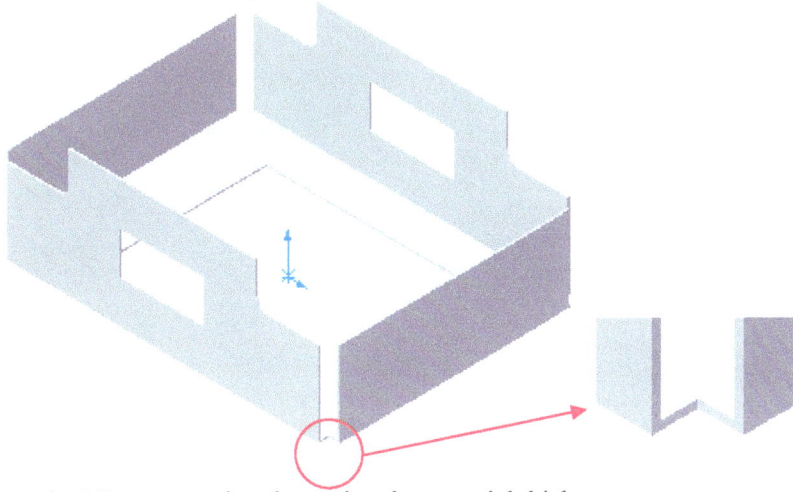

Use the **Measure** tool to determine the material thickness.

Pull down the "Tools" menu and pick **Measure**.

Select the left and right vertical edges of the end of the flange.

In the **Measure** dialog box, click on the **Units/Precision** icon.

The **Measure Units/Precision** dialog box appears.

Click on the **Use custom settings** radio button.

Then, change the **Decimal places** to '**4**'.

Click the **OK** button.

SolidWorks now shows that the **Distance** is **0.0480in**.

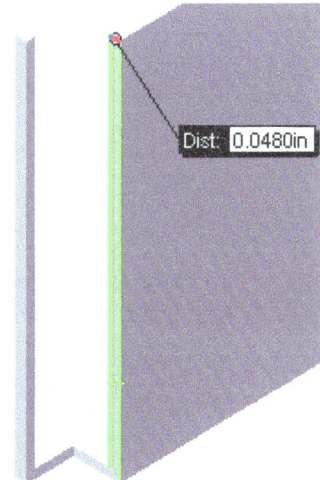

Press the **Escape** key and then the **Zoom to Fit** icon to display the whole part.

Convert the Part to Sheet Metal

Insert bends to the part by clicking the **Sheet Metal** icon in the control area of the CommandManager. Then, click the **Insert Bends** icon from the toolbar, or pull down the "Insert" menu and pick **Sheet Metal – Insert Bends**.

In the **Bends** PropertyManager, under **Bend Parameters**, set the **Bend Radius** to '**.05**'.

Under **Bend Allowance**, pick **Bend Table**. In the pull down, pick **SMG KFACTOR BASE BEND TABLE**.

In the graphics area, pick the top inside surface of the part to identify this as the **Fixed Face**.

Click the green check mark button at the top of the **Bends** PropertyManager to accept the settings and insert the bends.

Take a look at the FeatureManager design tree. There are four items of importance here: **Imported1**, **Sheet-Metal1**, **Flatten-Bends1**, and **Process-Bends1**.

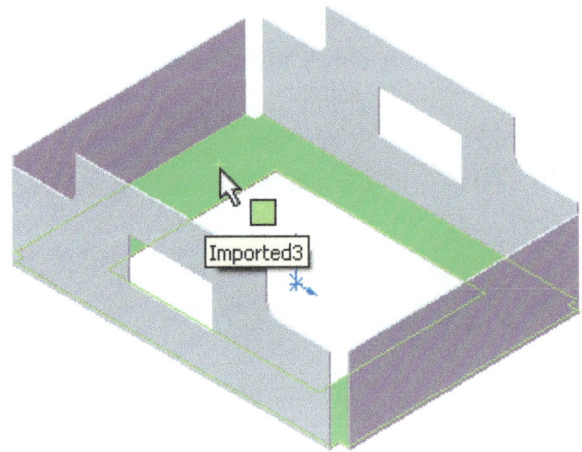

Right click on **Sheet-Metal1** and pick **Edit Feature**.

In the **Sheet-Metal1** PropertyManager, under **Bend Parameters**, the flange you selected is shown along with the default **Bend Radius** and the **Material Thickness** (which is not editable).

Under **Bend Allowance** is the method to be used, **Bend Table** in this case, and the name of the table you selected.

Click the green check mark button at the top of the **Sheet-Metal1** PropertyManager

In the FeatureManager design tree, click on the plus to the left of **Flatten-Bends1**. This expands the tree to show the four bend areas along with a sketch showing the sharp line for each bend area.

Click on each of the bends in the tree. As you select a bend area in the design tree, it highlights that bend in the model in the graphics area.

Right click on **Sharp-Bend3** and pick **Edit Feature**. This should be the bend area on the left side of the part with the cutout.

Uncheck **Use default radius** and enter the new radius value of '**.125**'.

Click the green check mark button at the top of the **Sharp-Bend3** PropertyManager.

Now right click on **Sharp-Bend4** and pick **Edit Feature**. This should be the bend area on the right side of the part with the cutout.

Again, uncheck **Use default radius** and enter the new radius value of '**.125**'.

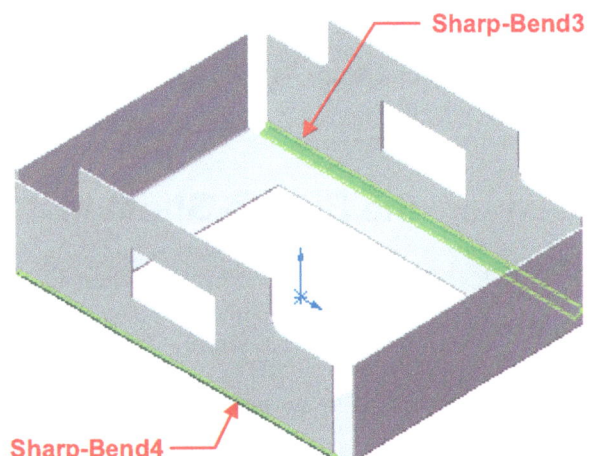

Click the green check mark button at the top of the **Sharp-Bend4** PropertyManager.

You don't need to change the **Bend Allowance** values since the k-factor table is being used. The values will be updated according to the table.

Make sure that the bends are valid by clicking the **Sheet Metal** icon in the control area of the CommandManager. Then, click the **Flatten** icon from the toolbar. You may also right click on **Flat-Pattern1** and pick **Unsuppress**.

In the bottom left corner of the graphics area, change the View orientation by clicking the pull down arrow and picking **Top**, or press **Ctrl+5**.

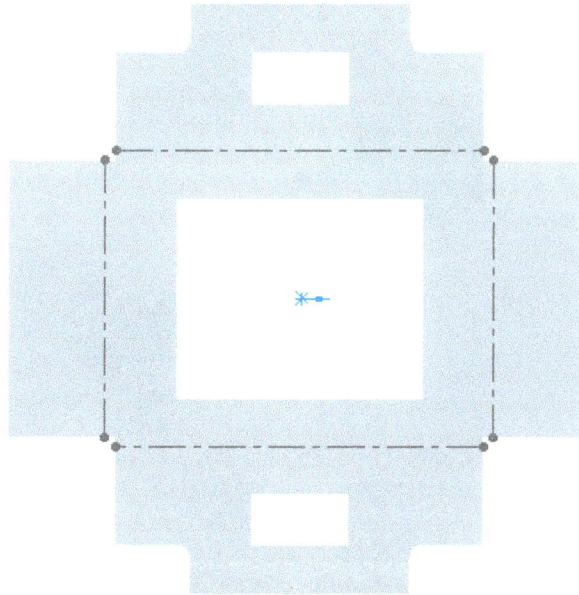

It looks good, but you can close those corners a bit. By the way, note that the **Closed Corner** command won't work here.

Click the **Flatten** icon from the toolbar to refold the part. You may also right click on **Flat-Pattern1** and pick **Suppress**.

In the bottom left corner of the graphics area, change the View orientation by clicking the pull down arrow and picking **Isometric**, or press **Ctrl+7**.

Close the Corners

Click the **Sheet Metal** icon in the control area of the CommandManager. Then, pick the **Base-Flange/Tab** icon from the toolbar, or pull down the "Insert" menu and pick **Sheet Metal – Base Flange**.

Select the front face of the part.

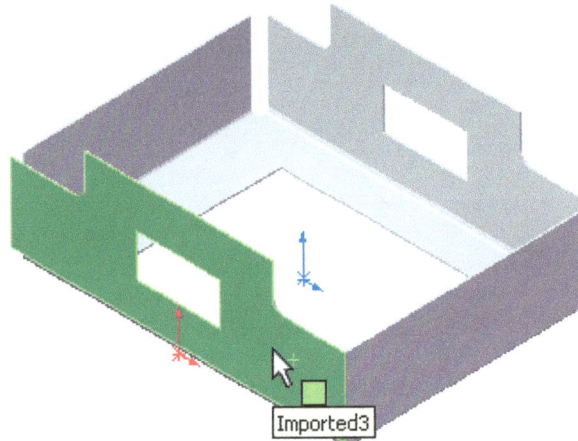

Change the display view to **Normal To** the sketch by picking **Normal To** in the View list in the lower left corner of the graphics area, or press **Ctrl+8**.

On the left side of the part, create a rectangle starting from the upper left corner of the flange to the left as shown using the **Rectangle** icon in the CommandManager, or pull down the "Tools" menu and pick **Sketch Entities – Rectangle**.

Ctrl select the left vertical line of the rectangle and the outside edge line of the left edge flange.

In the **Properties** PropertyManager, click the **Collinear** icon.

Drag the lower right corner of the rectangle until it snaps to the top of the bend line as shown.

Create a rectangle in the same way on the other end of the flange. Then, drag the lower left corner of the rectangle until it snaps to the top of the bend line as shown.

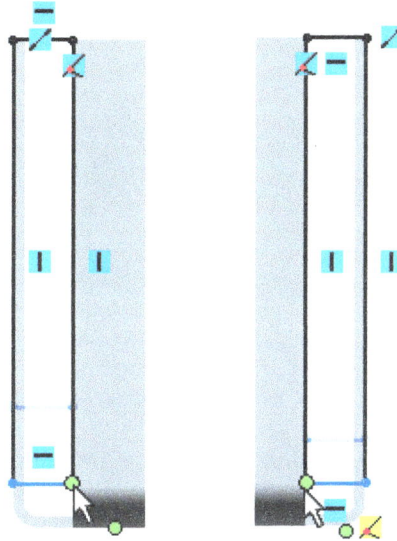

When both ends of the flange are complete, exit the sketch by clicking the **Exit Sketch** icon in the CommandManager or in the upper right corner of the graphics area.

In the bottom left corner of the graphics area, change the View orientation by clicking the pull down arrow and picking **Isometric**, or press **Ctrl+7**.

Click the **Base-Flange/Tab** icon from the toolbar, or pull down the "Insert" menu and pick **Sheet Metal – Base Flange**.

Select the right end of the part.

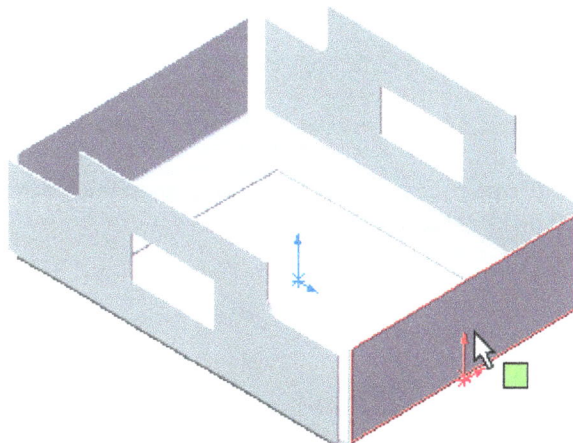

Change the display view to **Normal To** the sketch by picking **Normal To** in the View list in the lower left corner of the graphics area.

On the left side of the part, create a small rectangle starting from the upper left corner of the end flange to the left as shown using the **Rectangle** icon in the CommandManager, or pull down the "Tools" menu and pick **Sketch Entities – Rectangle**.

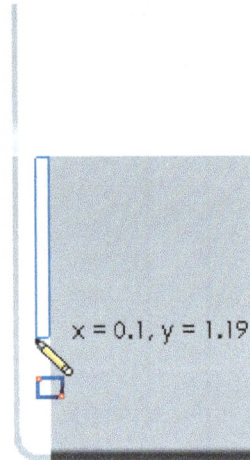

x = 0.1, y = 1.19

Add a '**.01**' horizontal dimension between the left vertical line of the rectangle and the inside edge of the front flange using the **Smart Dimension** icon in the CommandManager, or pull down the "Tools" menu and select **Dimensions – Smart**.

.01   .01

Do the same to the other side.

In the bottom left corner of the graphics area, change the View orientation by clicking the pull down arrow and picking **Isometric**, or press **Ctrl+7**.

Ctrl select the bottom lines of both rectangles and the bottom edge line of the front edge flange. You will have to zoom in and out to pick the correct lines.

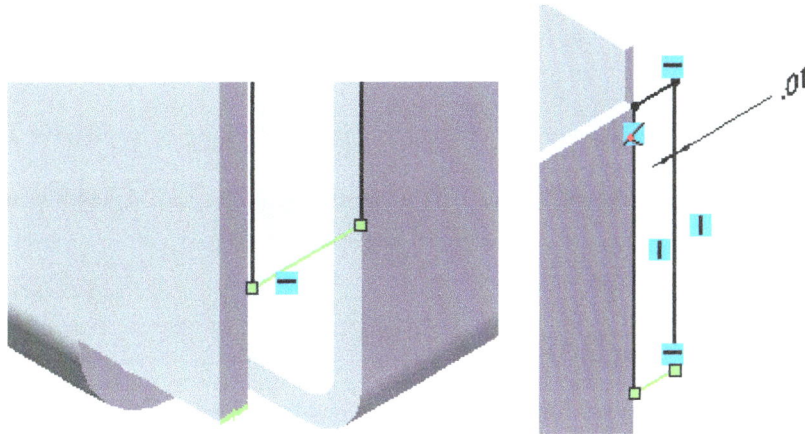

In the **Properties** PropertyManager, click the **Collinear** icon.

Exit the sketch by clicking the **Exit Sketch** icon in the CommandManager or in the upper right corner of the graphics area.

Click the **Zoom to Fit** icon to display the whole part.

Mirror the Tabs

Click the **Features** icon in the control area of the CommandManager. Then, click the **Mirror** icon in the CommandManager, or pull down the "Insert" menu and pick **Pattern/Mirror – Mirror**.

In the flyout FeatureManager design tree, select **Right Plane** as the **Mirror Face/Plane**.

Under **Features to Mirror**, select **Tab2** from the flyout FeatureManager design tree.

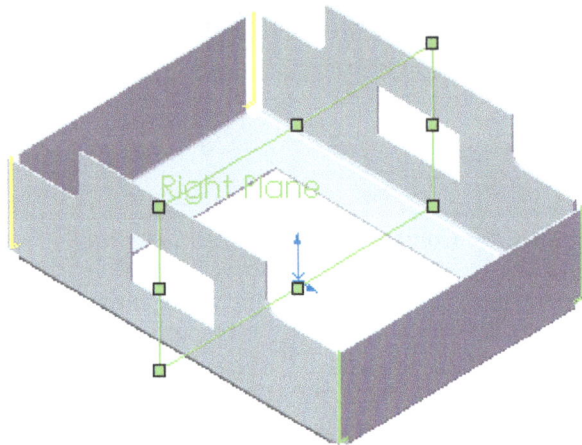

Click the green check mark button at the top of the **Mirror** PropertyManager.

Press **Escape**, and then, click the **Mirror** icon in the CommandManager, or pull down the "Insert" menu and pick **Pattern/Mirror – Mirror**.

In the flyout FeatureManager design tree, select **Front Plane** as the **Mirror Face/Plane**.

Under **Features to Mirror**, select **Tab1** from the flyout FeatureManager design tree.

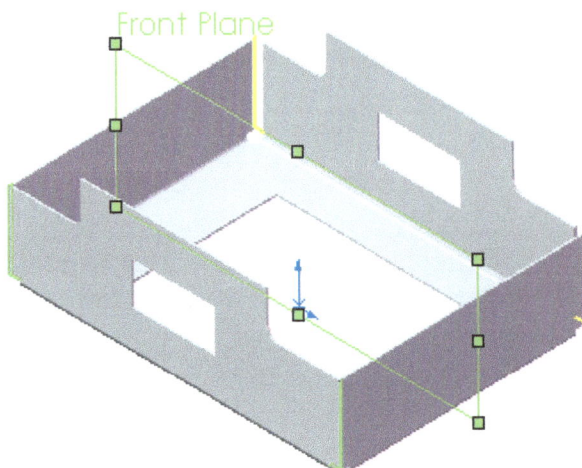

Click the green check mark button at the top of the **Mirror** PropertyManager.

In the bottom left corner of the graphics area, change the View orientation by clicking the pull down arrow and picking **Top**, or press **Ctrl+5**.

Take a look at how the flanges now meet in the corner areas.

Click the **Sheet Metal** icon in the control area of the CommandManager. Then, click the **Flatten** icon from the toolbar. You may also right click on **Flat-Pattern1** and pick **Unsuppress**.

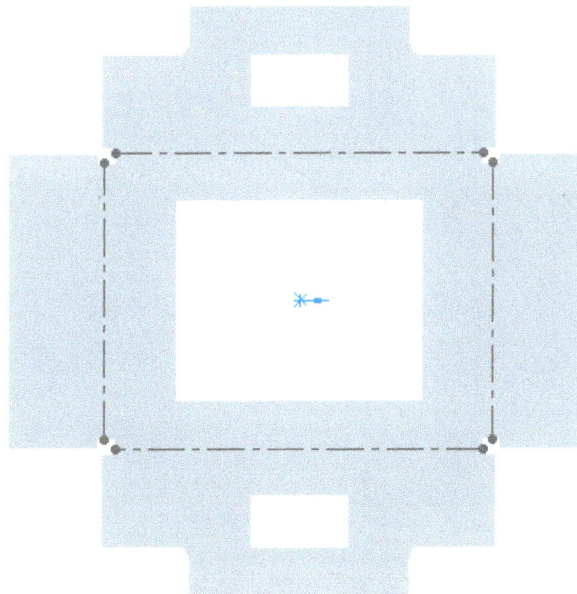

Click the **Flatten** icon in the CommandManager, or you may also right click on **Flat-Pattern1** and pick **Suppress**.

In the bottom left corner of the graphics area, change the View orientation by clicking the pull down arrow and picking **Isometric**, or press **Ctrl+7**.

Save the Part

Click the **Save** icon in the "Standard" toolbar, or pick **Save** from the "File" pull down menu.

The **Save As** dialog box appears. In the **File name** box, type '**BoxImport**' and click **Save**.

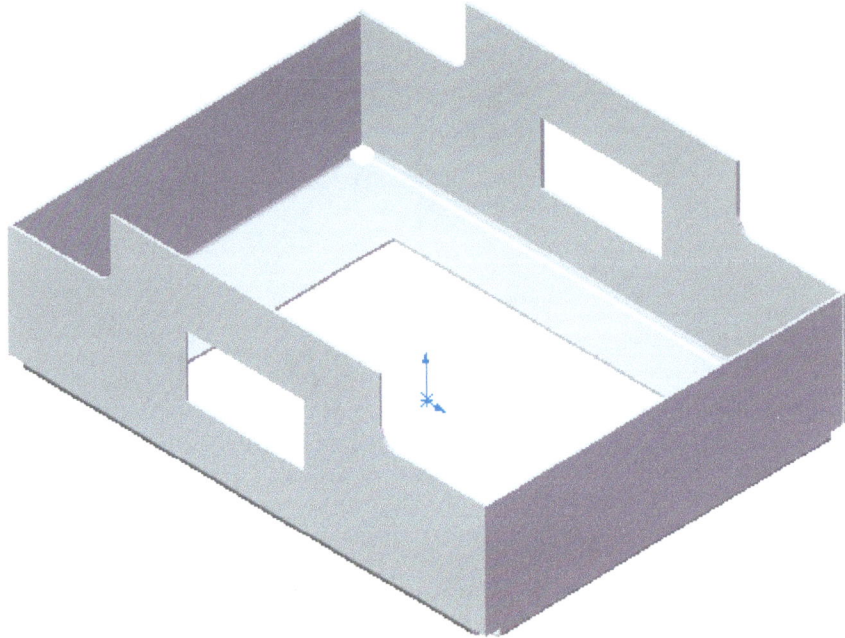

Chapter 13

Bend Sequencing with Configurations

Using the configuration feature of SolidWorks, you can create multiple models of the same part. This allows you to simulate the manufacturing process.

Yes, it is the beginning of creating the progressive die strip. However, we will get into that in Course 4, after you learn more about assemblies.

In the meantime, it is really cool to be able to see the part in the different phases of manufacture.

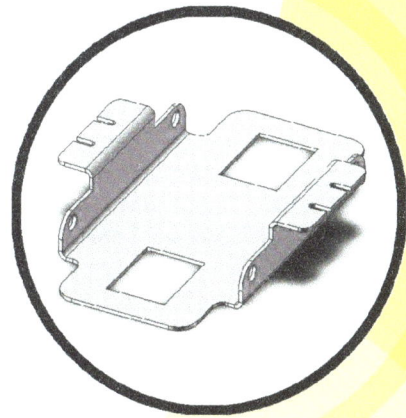

Create the Base Flange

Begin a new **Part** document by clicking the **New** icon in the "Standard" toolbar, or pull down the "File" menu and pick **New**.

In the FeatureManager design tree, right click on **Material**, and pick **Plain Carbon Steel** from the menu.

Create a base flange by clicking the **Sheet Metal** icon in the control area of the CommandManager. Then, click the **Base-Flange/Tab** icon from the toolbar, or pull down the "Insert" menu and pick **Sheet Metal – Base Flange**.

Select the **Top** plane when prompted to select a plane on which to sketch the feature cross-section.

Create the shape shown around the origin using the **Line** icon in the CommandManager, or pull down the "Tools" menu and pick **Sketch Entities – Line**.

Ctrl select the four small horizontal lines, and in the **Properties** PropertyManager, under **Add Relations**, click the **Equal** button.

Ctrl select the four small vertical lines, and in the **Properties** PropertyManager, click the **Equal** button.

Create a construction line diagonally across the middle of the shape as shown by clicking the **Centerline** icon in the CommandManager, or pull down the "Tools" menu and pick **Sketch Entities – Centerline**.

Select the inside top left corner and then the inside bottom right corner to create the centerline.

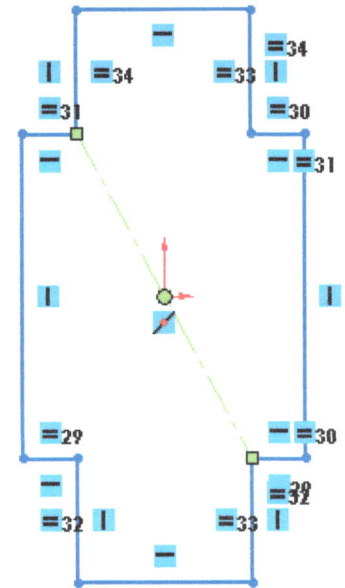

Press the **Escape** key to deselect the **Centerline** tool.

Ctrl select the diagonal line and the origin, and in the **Properties** PropertyManager, click the **Midpoint** button.

Add the dimensions as shown using the **Smart Dimension** icon in the CommandManager, or pull down the "Tools" menu and pick **Dimensions – Smart**.

Finally, add **.25** fillets to the outside corners and **.125** fillets to the inside corners as shown using the **Sketch Fillet** icon in the CommandManager, or pull down the "Tools" menu and pick **Sketch Tools – Fillet**.

In the SolidWorks dialog box stating that a segment being filleted has a midpoint or equal length relation, simply click **Yes** to continue.

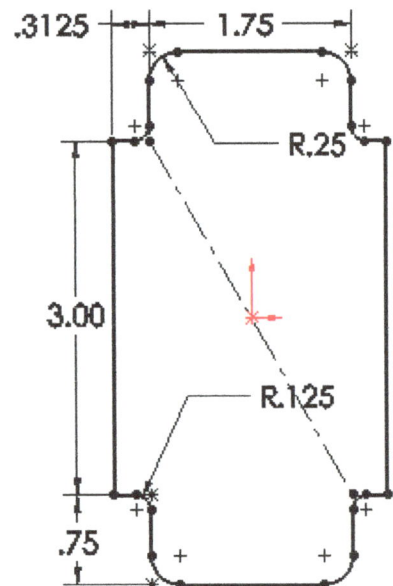

Exit the sketch by clicking the **Exit Sketch** icon in the CommandManager or in the upper right corner of the graphics area.

In the **Base Flange** PropertyManager, under **Sheet Metal Gauges**, check the **Use gauge table** check box.

Pick **CRS GAUGES** from the **Select Table** pull down.

Under **Sheet Metal Parameters**, set the **Thickness** to **18 Gauge**.

Make sure that the **Reverse direction** check box is checked.

Under **Bend Allowance**, set the **Bend Allowance Type** to **Bend Table**, and select **SMG KFACTOR BASE BEND TABLE** from the **Bend Table** pull down menu.

Click the green check mark button at the top of the **Base Flange** PropertyManager to accept the settings and create the part.

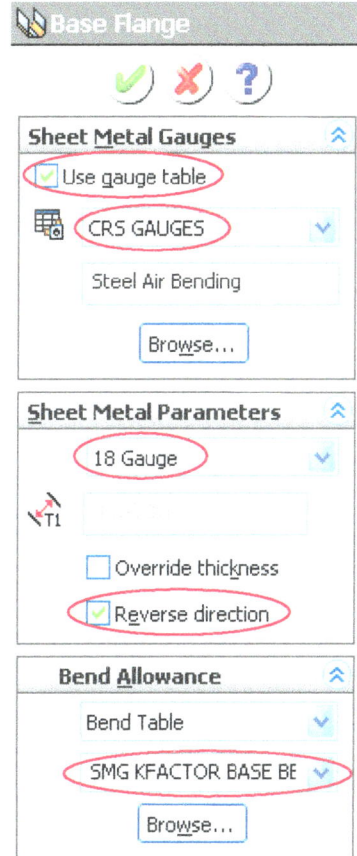

Add the Cutouts

Click the **Sheet Metal** icon in the control area of the CommandManager. Then, click the **Extruded-Cut** icon from the toolbar, or pull down the "Insert" menu and pick **Cut – Extrude**.

Select the top of the part.

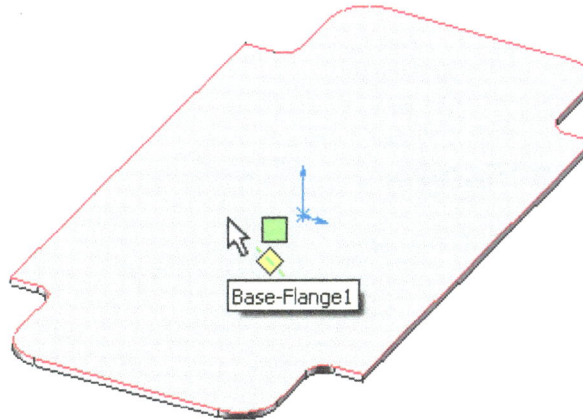

In the bottom left corner of the graphics area, change the View orientation by clicking the pull down arrow and picking **Top**, or press **Ctrl+5**.

Create a vertical centerline through the middle of the part using the **Centerline** icon in the CommandManager, or pull down the "Tools" menu and pick **Sketch Entities – Centerline**.

☐ Create a rectangle on the top half of the part and a rectangle on the bottom half of the part using the **Rectangle** icon in the CommandManager, or pull down the "Tools" menu and pick **Sketch Entities – Rectangle**.

= Ctrl select the four horizontal lines, and in the **Properties** PropertyManager, click the **Equal** button.

= Then, **Ctrl** select the four vertical lines, and in the **Properties** PropertyManager, click the **Equal** button.

∠ Right click on the top horizontal line of the top rectangle, and pick **Select Midpoint** from the menu. Hold down the **Ctrl** key and select the centerline. In the **Properties** PropertyManager, click the **Coincident** button.

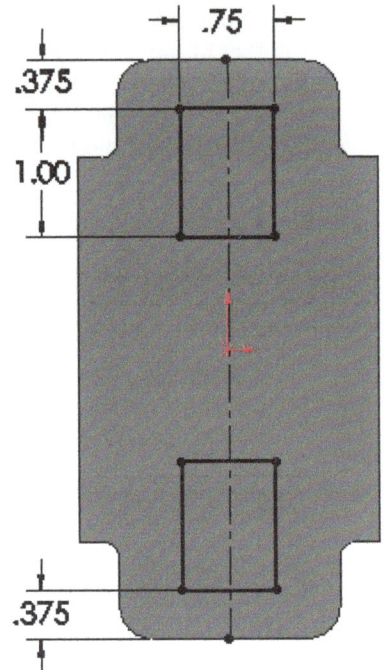

∠ Right click on the bottom horizontal line of the bottom rectangle, and pick **Select Midpoint** from the menu. Hold down the **Ctrl** key and select the centerline. In the **Properties** PropertyManager, click the **Coincident** button.

✎ Add the dimensions as shown using the **Smart Dimension** icon in the CommandManager, or pull down the "Tools" menu and pick **Dimensions – Smart**.

✏ Exit the sketch by clicking the **Exit Sketch** icon in the CommandManager or in the upper right corner of the graphics area.

In the **Cut-Extrude** PropertyManager, under **Direction1**, check **Link to Thickness**.

✅ Click the green check mark button at the top of the **Cut-Extrude** PropertyManager.

In the bottom left corner of the graphics area, change the View orientation by clicking the pull down arrow and picking **Trimetric**.

Create an Edge Flange

🐦 Click the **Edge Flange** icon in the CommandManager, or pull down the "Insert" menu and pick **Sheet Metal – Edge Flange**.

Select the left edge of the base flange and move the cursor up and click to set the direction of the flange.

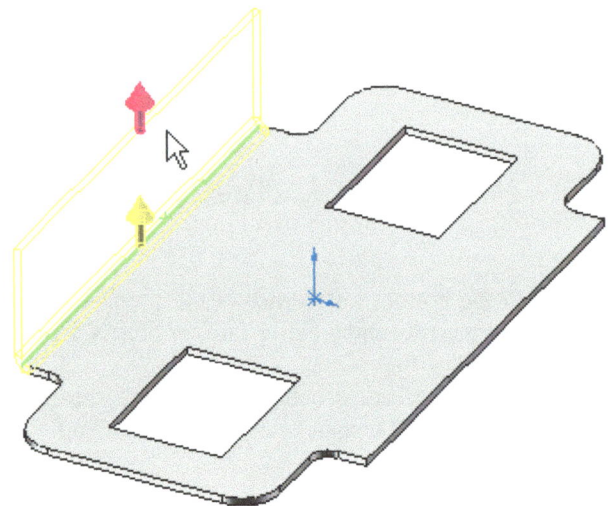

In the **Edge-Flange** PropertyManager, click on the **Edit Flange Profile** button.

In the bottom left corner of the graphics area, change the View orientation by clicking the pull down arrow and picking **Normal To**.

Edit the sketch by adding the additional lines and **.25** fillets as shown below.

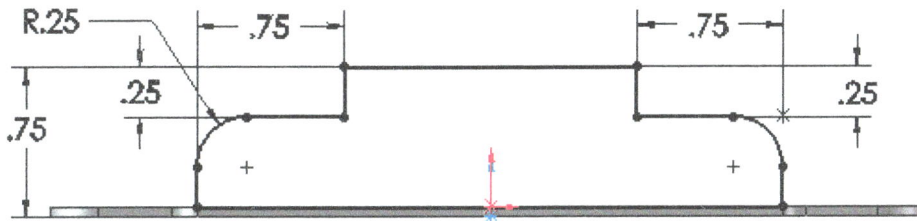

In the **Profile Sketch** dialog box, click the **Back** button.

In the **Edge-Flange** PropertyManager, click the **Material Inside Flange Position** button.

Click the green check mark button at the top of the **Edge-Flange** PropertyManager to accept the settings and create the flange.

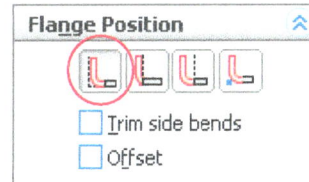

Add the Holes

Click the **Extruded-Cut** icon from the CommandManager, or pull down the "Insert" menu and pick **Cut-Extrude**.

Select the side of the part.

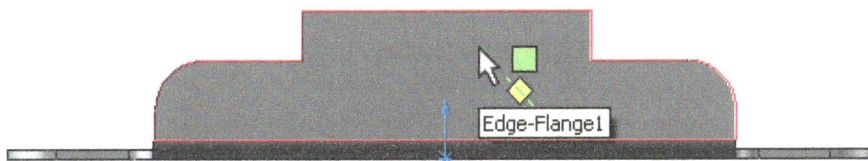

Create a vertical centerline through the middle of the part using the **Centerline** icon in the CommandManager, or pull down the "Tools" menu and pick **Sketch Entities – Centerline**.

Create a **.1875** circle as shown using the **Circle** icon in the CommandManager, or pull down the "Tools" menu and pick **Sketch Entities – Circle**.

Add the dimensions as shown using the **Smart Dimension** icon in the CommandManager, or pull down the "Tools" menu and pick **Dimensions – Smart**.

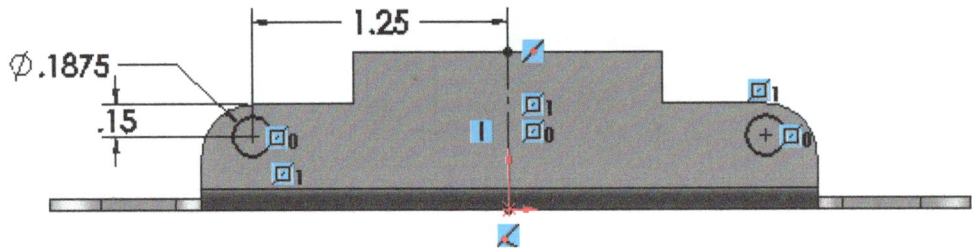

Ctrl select the circle and the centerline, and click on the **Mirror Entities** icon in the CommandManager, or pull down the "Tools" menu and pick **Sketch Tools – Mirror**.

Exit the sketch by clicking the **Exit Sketch** icon in the CommandManager or in the upper right corner of the graphics area.

In the **Cut-Extrude** PropertyManager, under **Direction1**, check **Link to Thickness**.

Click the green check mark button at the top of the **Cut-Extrude** PropertyManager.

In the bottom left corner of the graphics area, change the View orientation by clicking the pull down arrow and picking **Trimetric**.

Create Another Edge Flange

Click the **Edge Flange** icon in the CommandManager, or pull down the "Insert" menu and pick **Sheet Metal – Edge Flange**.

Select the top outside edge of the left edge flange and move the cursor to the left and click to set the direction of the flange.

Pick here

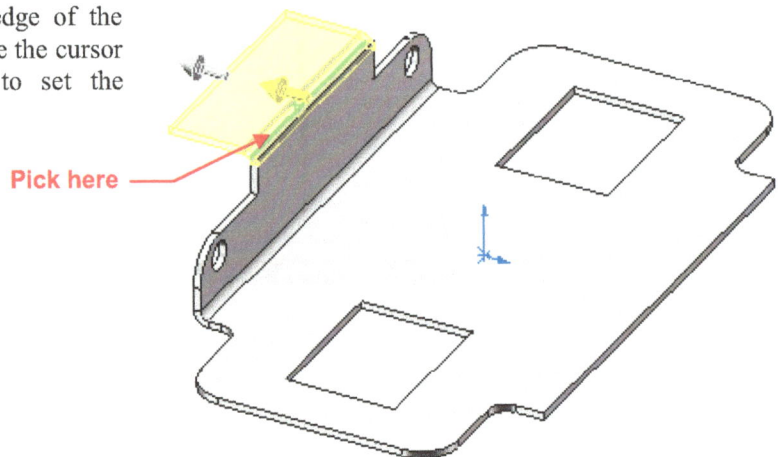

In the **Edge-Flange** PropertyManager, click on the **Edit Flange Profile** button.

In the bottom left corner of the graphics area, change the View orientation by clicking the pull down arrow and picking **Normal To**. If the part is now being viewed from the bottom, change the View orientation by clicking the pull down arrow and picking **Top**.

Edit the sketch by first adding the '**0.50**' dimension on the side of the part.

Next, add **.25** fillets to the outside corners as shown using the **Sketch Fillet** icon in the CommandManager, or pull down the "Tools" menu and pick **Sketch Tools – Fillet**.

Create the slot on the left side and then mirror it about the center line.

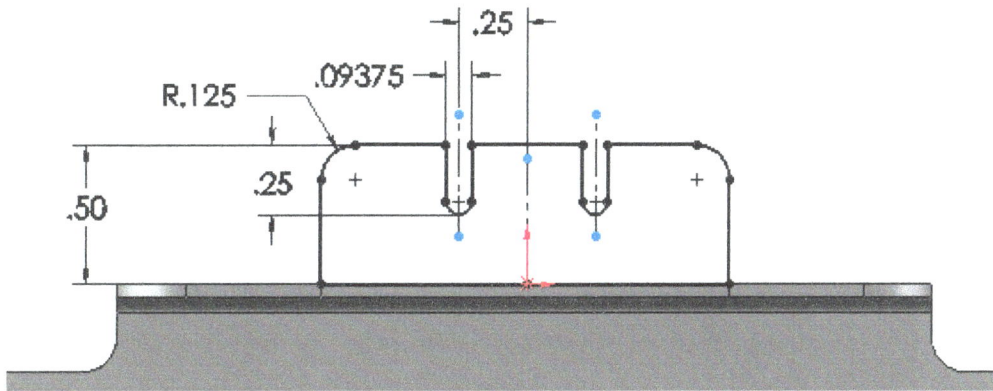

In the **Profile Sketch** dialog box, click the **Back** button.

In the **Edge-Flange** PropertyManager, click the **Material Inside Flange Position** button.

Click the green check mark button at the top of the **Edge-Flange** PropertyManager to accept the settings and create the flange.

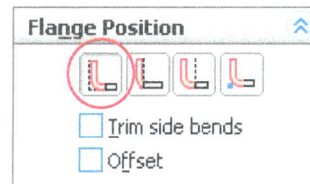

In the bottom left corner of the graphics area, change the View orientation by clicking the pull down arrow and picking **Trimetric**.

Mirror the Edge Flanges

Click the **Features** icon in the control area of the CommandManager. Then, click the **Mirror** icon from the toolbar, or pull down the "Insert" menu and pick **Pattern/Mirror – Mirror**.

In the flyout FeatureManager design tree, select **Right Plane** as the **Mirror Face/Plane**.

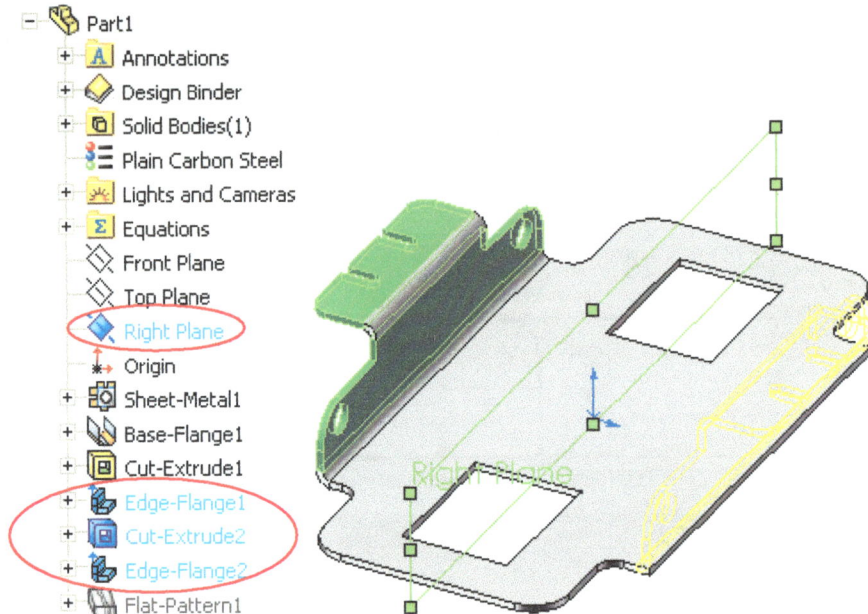

Under **Features to Mirror**, select **Edge-Flange1**, **Edge-Flange2**, and **Cut-Extrude2** from the flyout FeatureManager design tree.

Click the green check mark button at the top of the **Mirror** PropertyManager.

Save the Part

Click the **Save** icon in the "Standard" toolbar, or pick **Save** from the "File" pull down menu.

The **Save As** dialog box appears. In the **File name** box, type '**Mounting Bracket**' and click **Save**.

Add Configurations

In the FeatureManager Tree area, click on the **ConfigurationManager** tab. You can resize the panel at any time.

In the ConfigurationManager, right click on the part name (Mounting Bracket Configuration) and pick **Add Configuration**.

In the **Add Configuration** PropertManager, click in the **Configuration name** box and type 'Step 1'.

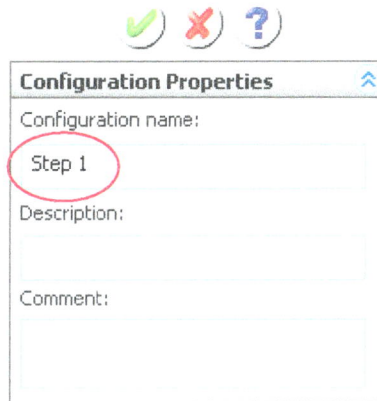

Click the green check mark button at the top of the **Add Configuration** PropertyManager.

Click the **Sheet Metal** icon in the control area of the CommandManager. Then, click the **Flatten** icon from the toolbar. You may also right click on **Flat-Pattern1** in the FeatureManager design tree and pick **Unsuppress**. This displays the flat pattern as the current state of the **Step 1** configuration.

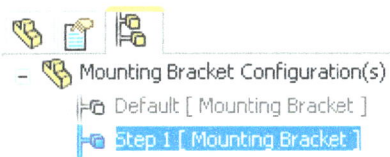

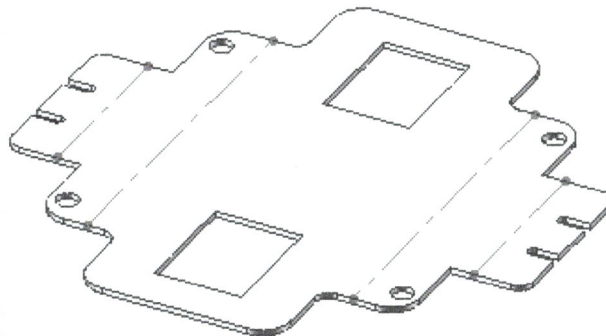

You can make additional copies of the **Step 1** configuration to show the sequence of operations. To do this, in the ConfigurationManager, click on **Step 1** to highlight it.

Then, press **Ctrl+C**, or pull down the "Edit" menu and pick **Copy**.

Finally, press **Ctrl+V**, or pull down the "Edit" menu and pick **Paste**. Paste three copies of **Step 1**.

Step 1 Configuration

To further define the **Step 1** configuration, click the **Features** icon in the control area of the CommandManager. Then, click the **Extruded Boss/Base** icon from the toolbar, or pull down the "Insert" and pick **Boss/Base – Extrude**.

Select the top of the flattened part.

In the bottom left corner of the graphics area, change the View orientation by clicking the pull down arrow and picking **Normal To**.

Create a rectangle around the part using the **Rectangle** icon in the CommandManager, or pull down the "Tools" menu and pick **Sketch Entities – Rectangle**.

Add the **.25** dimensions as shown using the **Smart Dimension** icon in the CommandManager, or pull down the "Tools" menu and pick **Dimensions – Smart**.

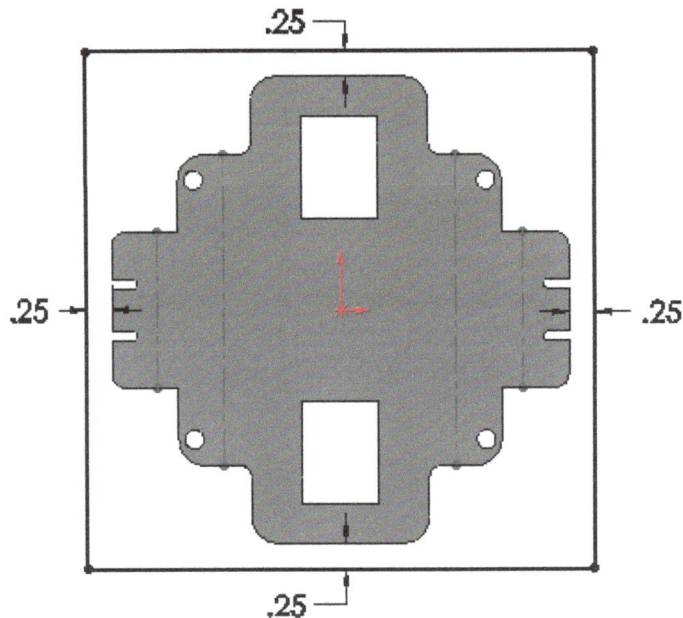

Exit the sketch by clicking the **Exit Sketch** icon in the CommandManager or in the upper right corner of the graphics area.

In the bottom left corner of the graphics area, change the View orientation by clicking the pull down arrow and picking **Trimetric**.

In the **Extrude** PropertyManager, under **Direction 1**, click on the **Reverse Direction** button next to the **End Condition**.

Then, check **Link to Thickness**, and click the green check mark button at the top of the **Extrude** PropertyManager.

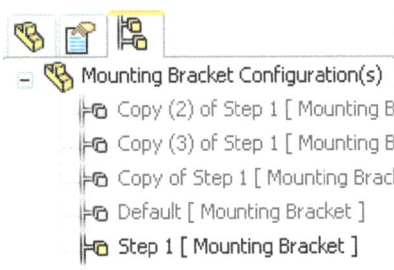

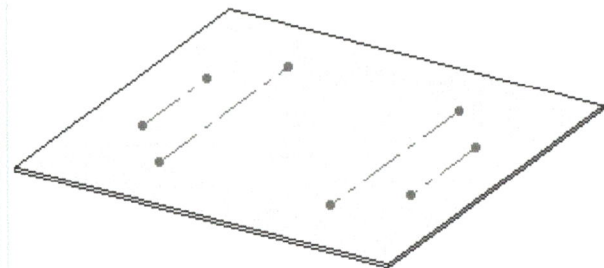

Step 2 Configuration

Show the **Copy of Step 1** configuration by double clicking on **Copy of Step 1** in the ConfigurationManager, or right click on the **Copy of Step 1** configuration and pick **Show Configuration**.

Rename the configuration by clicking on **Copy of Step 1** once, pause and clicking on it again (slow double clicking) to activate the the text. Change the name to '**Step 2**'.

In the FeatureManager Tree area, click on the **Feature Manager design tree** tab.

In the Feature Manager design tree, right click on **Cut-Extrude1** and pick **Suppress**.

Right click on **Cut-Extrude2** and pick **Suppress**.

In the FeatureManager Tree area, click on the **ConfigurationManager** tab.

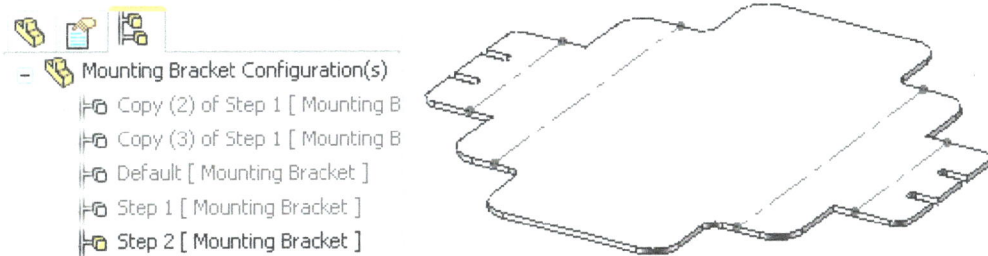

Step 3 Configuration

Show the **Copy (2) of Step 1** configuration by double clicking on **Copy (2) of Step 1** in the ConfigurationManager, or right click on the **Copy (2) of Step 1** configuration and pick **Show Configuration**.

Rename the configuration by clicking on **Copy (2) of Step 1** once, pause and clicking on it again (slow double clicking) to activate the the text. Change it to '**Step 3**'.

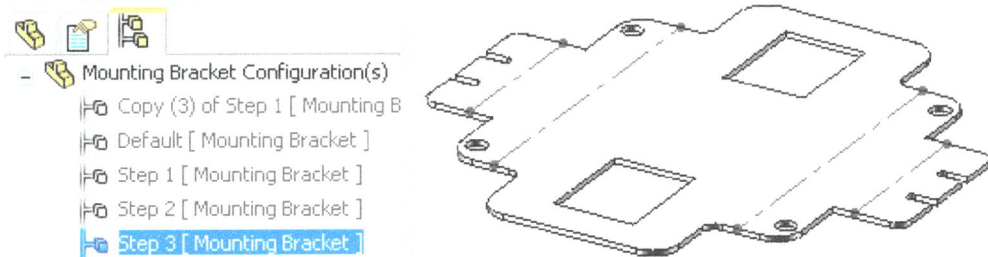

Step 4 Configuration

Show the **Copy (3) of Step 1** configuration by double clicking on **Copy (3) of Step 1** in the ConfigurationManager, or right click on the **Copy (3) of Step 1** configuration and pick **Show Configuration**.

Rename the configuration by clicking on **Copy (3) of Step 1** once, pause and clicking on it again (slow double clicking) to activate the the text. Change it to 'Step 4'.

You can split the ConfigurationManager and either display two ConfigurationManager instances, or combine the ConfigurationManager with any of the other tabs.

To do this, drag the top bar of the FeatureManager design tree down.

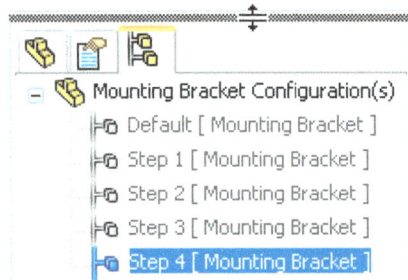

- Mounting Bracket Configuration(s)
  - Default [Mounting Bracket]
  - Step 1 [Mounting Bracket]
  - Step 2 [Mounting Bracket]
  - Step 3 [Mounting Bracket]
  - Step 4 [Mounting Bracket]

This way, you can easily show different configurations and compare what features are suppressed in the FeatureManager design tree.

In the FeatureManager design tree, click on the plus sign next to **Flat-Pattern1**.

Right click on **Flatten<Edge-Bend2>1** and pick **Suppress**.

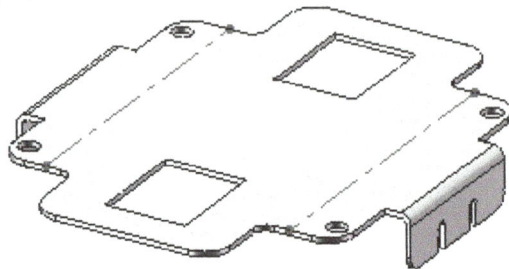

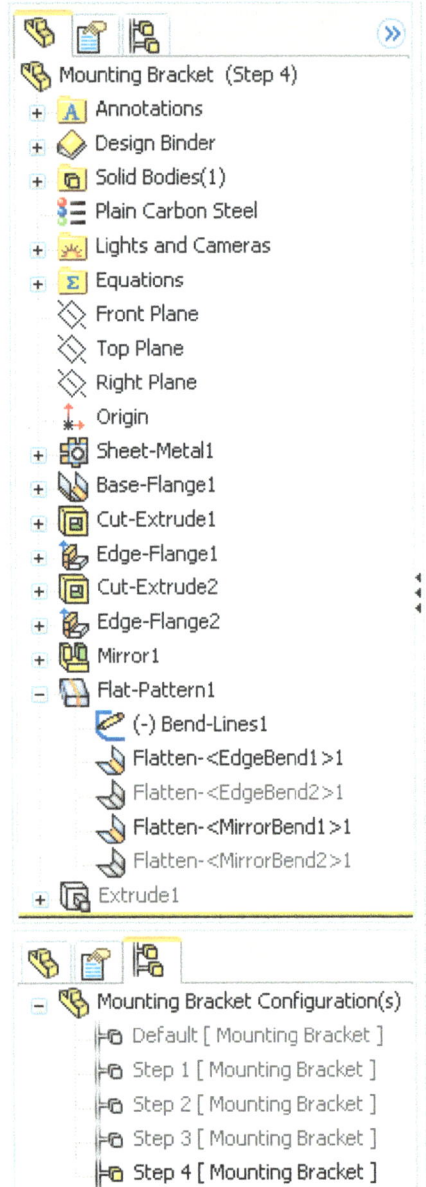

Right click on **Flatten<Mirror-Bend2>1** and pick **Suppress**.

Mounting Bracket (Step 4)
- Annotations
- Design Binder
- Solid Bodies(1)
- Plain Carbon Steel
- Lights and Cameras
- Equations
- Front Plane
- Top Plane
- Right Plane
- Origin
- Sheet-Metal1
- Base-Flange1
- Cut-Extrude1
- Edge-Flange1
- Cut-Extrude2
- Edge-Flange2
- Mirror1
- Flat-Pattern1
  - (-) Bend-Lines1
  - Flatten-<EdgeBend1>1
  - Flatten-<EdgeBend2>1
  - Flatten-<MirrorBend1>1
  - Flatten-<MirrorBend2>1
- Extrude1

- Mounting Bracket Configuration(s)
  - Default [Mounting Bracket]
  - Step 1 [Mounting Bracket]
  - Step 2 [Mounting Bracket]
  - Step 3 [Mounting Bracket]
  - Step 4 [Mounting Bracket]

Step 5 Configuration

Show the **Default** configuration by double clicking on **Default** in the ConfigurationManager, or right click on the **Default** configuration and pick **Show Configuration**.

Rename the configuration by clicking on **Default** once, pause and clicking on it again (slow double clicking) to activate the the text. Change it to '**Step 5**'.

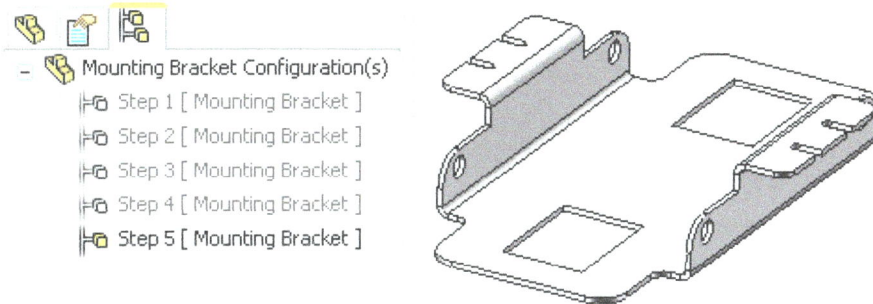

You now have five different configurations of the part. These will all be saved together when you save the part.

Save the Part

Click the **Save** icon in the "Standard" toolbar, or pick **Save** from the "File" pull down menu.

Course 4 will show you how to put these configurations together to simulate the die strip.

Congratulations

You have now completed Course 3. Unfolding your parts is easy. Getting the correct flat pattern is up to you. SolidWorks provides the tools for you to specify the bend parameters for the pat as a whole or for each individual bend area. It is up to you to specify how you want them applied.

For ease of use and consistency, it is recommended that you create a set of tables based on your existing values in use by your organization. While the k-factor method is the most accurate overall, many people seem to be afraid of it. Don't be! Take some time to learn it and apply it. Once you do, your unfolding process will become easier and more consistent.

"SolidWorks for the Sheet Metal Guy – Course 4: Complex Parts" will introduce you to some more advanced methods. Transitions and cones will be demonstrated. Learn to create assemblies and to edit parts within the assembly. Also, see how to put together the part configurations to model the die strip.

SolidWorks for the Sheet Metal Guy

Appendices

SolidWorks for the Sheet Metal Guy

Appendix A

SolidWorks allows you to create three different styles of Bend Tables. They are Bend Allowance, Bend Deduction, and k-factor.

Whenever possible, I recommend you use the k-factor table. It allows you to enter a range of data and then interpolates any other data needed for your parts. While the other types of files also interpolate the in between data, you need to find the k-factor in order to come up with the values needed for these other tables. So why do all that work, once you have the k-factor, just use it.

To create a new table, simply open one of the existing tables in Excel and immediately do a **Save as** to rename it.

If you wish to create the table directly in SolidWorks, you can pull down the **Insert** menu and select **Sheet Metal**, **Bend table**, **New**. A dialog will appear allowing you to specify the Units, Table Type, and the File name.

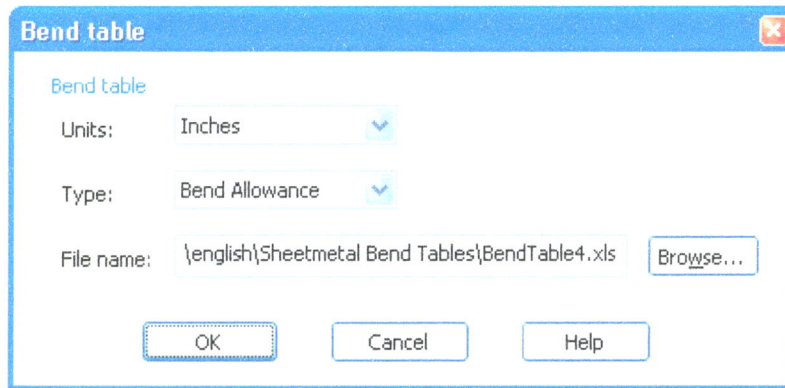

Select the desired units, Table Type, and enter the name for the table. Use the **Browse** button to make certain which folder you are placing the file in. Then, click the **OK** button and an Excel spread sheet opens for you with blank data fields.

To close the table, you simply click the cursor anywhere on the screen, outside of the table window. Once you close the table, you will need to use the **Edit** menu and pick **Bend Table – Edit Table** to reopen the table.

Either way, the basic table has three header lines: **Units**, **Type**, and **Material**.

The **Units** statement specifies the units for the values in the table. You may specify the units as: millimeters, centimeters, meters, inches, or feet. As you will note in the comment to the right of this field, you must spell out the units. Do not abbreviate.

The **Type** field tells SolidWorks whether this is a table of Bend Allowance, Bend Deduction, or K-Factor. Again spell it the way it appears in the comment to the right.

The **Material** field is really just text for your reference. You may enter any name or material type you wish as a reference.

K-factor tables are discussed in Chapter 2. For Bend Allowance and Bend Deduction tables, the rest of the worksheet is a series of charts based on material thickness. The vertical column **A** lists the angles of bend and a horizontal row lists the radii.

The default tables use fractions for the radii and material thickness values. You will probably want to change them to be in decimal format and show more than two decimal places.

	A	B	C	D	E	F	G	H	I	J	K	L
1												
2	Unit:	Inches			#	Available Units:	Millimete	Centimet	Meters	Inches		Feet
3	Type:	Bend Deduction			#	Available Types:	Bend Allowance		Bend Deduction			K-Factor
4	Material:	304 Stainless Steel										
5	#											
6												
7	Thickness:	0.0360										
8	Angle						Radius					
9			0.0156	0.0313	0.0469	0.0625	0.0938	0.1250	0.1875	0.2500	0.3125	0.3750	0.5000
10	15											
11	30											
12	45											
13	60											
14	90											
15	135											
16	180											

By default there are only two charts in the file. Before you cut and paste these to make more, edit the **Radius** values to be only the radii you use in your manufacturing process. Don't eliminate all of them and insist you only use one radius. Think of all the different parts you make and come up with a real list of radii. The **Gauge Table** should already contain this information.

Now look down the list of bend angles. Do you need all of these? Again, don't just delete a bunch. Adding the extra data will help SolidWorks when it has to interpolate a value for you. List all of the angles that you regularly bend. But don't forget that 180 degrees is probably the angle of the **Hem** feature and may require a special radius value to go along with it.

Now that you updated the **Angle** and **Radius** values, copy the chart as necessary to make one for each material thickness you manufacture from. Open the material gauge table you created and update the **Thickness** values with your standard material thicknesses.

You are now ready to enter the real data into the table. Start with the row for the 90 degree angle, since this will be the easiest. From there, work through each item of each chart entering the data. The **Bend Calculator** on the Sheet Metal Guy website (www.SheetMetalGuy.com) will be a big help for doing this.

Since different material types can have different thicknesses, such as galvanized versus cold rolled steel, you may need to create a separate table for each of these. If you are using air bending and bottom bending techniques, you may need separate tables for this reason as well.

This is a lot of work to get started, but once you have done it and tested the results, it should be good forever. Or at least until you change your manufacturing techniques.

Appendix B

Where do you get the values to go in here? Good question! Most people have a chart hanging on the wall that tells them the values to use. Bad news, the chart is typically only for 90 degrees. But there are tools out there to help you. You can use the bend calculator on the Sheet Metal Guy website (www.SheetMetalGuy.com) to find the values that you need to complete these tables. Your chart on the wall may be bend deduction or bend allowance. The industry switches definitions of these two terms frequently.

To find the **Bend Calculator**, open your Internet Explorer and go to http://www.sheetmetalguy.com. In the navigation bar you will find **Valuable Tools**. The fly out menu includes the **Bend Calculator**.

You will need to know the definitions and formulas to get the job done right.

Bend Allowance is the length through the arc at the neutral axis. The formula for the circumference of a circle is 2 times the radius times PI (3.14). So, to get the length of the arc of your bend, also multiply by the angle divided by 360. If you simplify the formula, it looks like this:

$$BA = radius * PI * angle / 180$$

The radius of your bend is from the inside. You need to get the radius to the neutral axis. To do this, you must add a percentage of the material thickness to the inside bend radius. This percentage is called the k-factor. See, even to calculate the bend allowance, you need to know the k-factor. This is why it is recommended to always use the k-factor method.

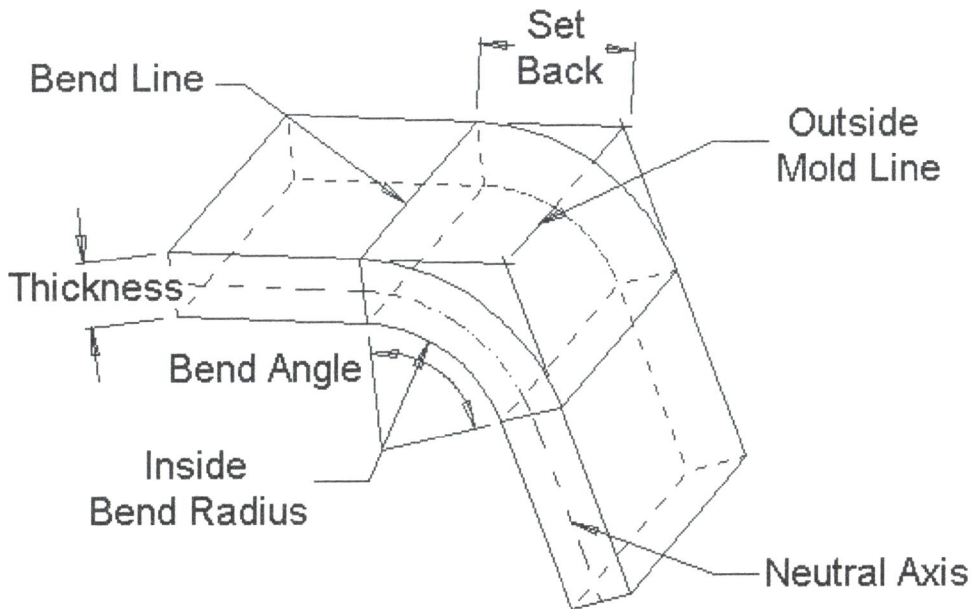

The k-factor is normally a value between **0** and **0.5**. After all, the neutral axis logically is at the middle of the material thickness. But, since the metal doesn't allow the inside surface to properly compress when you bend it, the neutral axis moves toward the inside. The smaller the radius, the farther it moves, in theory. There are also times when you will need to use a k-factor outside of this range in order to get the correct results, due to the model having the incorrect geometry.

The **Bend Calculator** will allow you to determine the k-factor used to create the chart you may have laying around the office.

Bend Compensation is the adjustment amount. If you squared off the bend area, removing the arcs, Bend Compensation is the distance from the start of the bend area to the sharp corner, minus the Bend Allowance. For example, on a 90 degree bend of 0.048 material with a 0.0625 inside radius, the Bend Compensation is equal to 2 (for two sides of the bend area) * (0.0625 + 0.048) – 1.57 * (0.0625 + 0.048 * 0.38) = 0.0942382.

The first part of this formula is what is called Setback. Setback is the distance from the start of the bend area to the sharp corner. The real formula based on measuring the outside of the bend area is:

$$\text{Setback} = \tan(\text{Angle} / 2) * (\text{Radius} + \text{Thickness})$$

It must be noted that when the bend angle is 180 or greater, the Setback does not exist since the intersection of the flanges does not exist or is on the wrong side of the bend area.

Now that you have the basics, you can do your own math. Or better yet, just use the **Bend Calculator**. It is so much easier.

Index